# THE DOMINIUM

HASANUDDIN

# THE DOMINIUM

Sequencing antimatter
and gravity effects:
Big Bang to black hole;
and implications for
a manmade near-future
doomsday:

End-of-all-life on Earth

Editorial assistance: Judy Feldmann

Library of Congress designation pending

Bishmillahrhakhmanirrakhhimm....

Kepada dunia indah dan semua manusianya

S T A R T L I N G

S T A R T I N G

S T A R I N G

S T R I N G

S T I N G

S I N G

S I N

I N

I

.

# CONTENTS

# INTRODUCTION

Dr Suess's classic story *Horton Hears a Who* chronicles the efforts of a lovable elephant trying to save the inhabitants of a mini-world on the brink of calamity. The world exists on a speck of dust so tiny that it is overlooked and unnoticed by all those living within Horton's world—except Horton. He is the only one aware of the danger because of his sensitive hearing: listening to their pleas, he strives to save the little planet.

To the Who's, Horton's world doesn't exist. Their realities are contained within the confined area of their speck of dust. They live their lives, go to work, play, and conduct science solely under the assumption that their world constitutes all that is worth knowing. The parallels are strong between this fictional children's story and the work of non-fiction hard-science that is about to follow.

Like the Who's, we on Earth tend to consider our little planet to be the full extent of all that is worth knowing. Many among us are even more nearsighted than that, caring only about events and happenings within our own little countries, towns, or neighborhoods. There is no crime in being simple and content, like the Who's; but being content does not prevent outside events from reaching into our little microworlds, potentially leading to ultimate destruction.

Unlike the Who's, scientists among us know quite a good deal about the extent of the Universe outside our little speck of dust. We know that we live within a solar system with a star named Helios at the center. We know that Helios exists within a larger conglomeration of millions stars that form a disk rotating around a central point (thought to be a massive black-hole). This conglomeration of stars we call the Milky Way. We have been amazed by pictures from the Hubble Space Telescope showing millions upon millions of other galaxies, seemingly evenly distributed into the far reaches of the sky. The picture used for the cover of this book is such an example of the extent of the universe. The genuine stars shown in this picture, like the large white blotch on the edge, have hard "compass" points of light, due to lensing effects and overexposure. Every other speck and smudge in the cover photo is actually a galaxy, each comprised of hundreds of millions of stars.

We do not live on a dust speck floating aimlessly in the cosmos, like the world of the Who's; our planet, our solar system, and our galaxy do have some predictability and structure. However—also unlike the Who's—we do know of real dangers that exist within the Universe. We have observed the activities of black-holes, strange forms of matter that act like giant trash-compactors that consume, compact, and devour adjacent material. We have watched the phases of distant stars as they age, run out of fuel, and in some cases, when they become critically imbalanced, implode to form a new black-hole, spewing dust far into space. We know that our own little planet has been bombarded with rocks and debris, suffering such random collisions throughout its history. Mass extinctions have even been attributed to the effects of past collisions. There are even some who fear the apocalyptic consequences of another such collision. Fortunately, there are no known sizable space objects that appear to have a significant chance of striking the Earth any time soon. And there are no known or theorized black-holes within sucking distance of our solar system.

The cosmos is real, yet our puny status makes travel outside of the Earth nearly impossible. And in that sense, we are very much like the Who's. We cannot leave our little microworld, so we care little about what is going on outside of it. Although there are scientists peering toward the outside, the majority are content to go about business, live, die, and play without being bothered by the big questions of where it all came from, how it works currently, or to where it is all going. Humankind has evolved and has lived without knowing the answers to these questions up until now. There has always been the errant philosopher or scientist who dreams of possible answers to these questions, while the majority neither care nor want to know the answers.

Up until now, that was just fine. But the Earth is currently under a very real threat. The danger does not come from outside our planet; rather the danger is synthetic/man-made. Even so, annihilation would be both final and complete. The scenario involves man's eternal quest to understand the unknown and further the achievements of science, wrapped in the arrogance and egos of those who think they know more and are better than the rest of humanity. Actually, the current momentum that is leading the world to its end would make for a very interesting and thrilling science-fiction story...if it weren't so very real.

Unfortunately, this book is not a work of fiction. There are several laboratories currently racing to be the first to create and observe black-hole material. Scientists are blinded by their professional lust to attain the coveted Nobel Prize. "Surely," they reason, "whoever is the first to create this 'new' form of matter will be guaranteed the Nobel."

In defense of the scientist, there are currently many conflicting theories surrounding black-holes. The common assumption is that black-holes require vast amounts of mass in order to attain stability. This assumption is backed up by observations of aging stars: only the very largest are observed to go super-nova and become black-holes. However, this fact does not preclude the formation of black-hole material via synthetic means. Hence the quest and race to create the first man-made black-hole. What the common assumption gives is a way to reassure the public (and one's self) that black-hole material synthesized in the lab will not be stable. To the top physicist contemplating the Nobel, this reassurance is necessary to continue working. The assistants, interns, technicians, and bureaucrats supporting current research have put their faith in these few top physicists. "Surely nothing will go wrong," the surrounding supporters reason, "because people much smarter than myself have thought these things through. We are all doing something noble...and possibly Nobel."

The argument that vast quantities of matter are needed to sustain a stable black-hole is tossed aside by the scientific model revealed in this book. The Dominium model predicts the existence of black-holes (and antimatter black-holes) of all sizes. The Dominium model also neatly explains all celestial events, starting with the Big Bang, extending through modern times, and ending with a description of how the next Big Bang will occur. The model is clean, and free of the contradictions of experimental or observational anomalies that confound other theories or models. Phenomena that lacked explanation in other models (dark-matter, solar wind, the central galactic black-hole, and the absence of antimatter on Earth) are all predicted by the Dominium. Because of the beauty, completeness, and thoroughness of the Dominium model, it needs to be taken seriously in future/current decision-making choices. That is: current attempts to create black-hole material need to be stopped.

This work is being delivered through an exceptionally common and insignificant man. True, the author has worked and received training at CERN, NASA, and MIT; but, he is not a PhD, or a top physicist, nor does he aspire to win the Nobel. This book was not written to gain fame, money, or career advancement. How this book was written is actually stranger than the words, theories, and warnings that it delivers. The model actually arose from a series of visions and epiphanies that can only be attributed to divine origin. But regardless of their origin, the model is sound. To verify its completeness, the author has combed current scientific journals to cross-reference and support all major premises with modern research and understanding.

Early versions of the model have been shared with top scientists and theorists at MIT, NASA, and CERN. None of these men could find fault with the science of the Dominium. One theorist at CERN, however, was enraged that an outsider dared question the work of hundreds of people (including himself) and the assumptions of Einstein. Although this man's rage was passionate, neither of his arguments is sound. Einstein and those "hundreds of people" are/were just flawed human beings. None of them was ever infallible; no theory is correct beyond any scientific doubt. The author of this book is also flawed. However, the author's own personal character defects and past mistakes does not invalidate the science, logic, or soundness of the delivered Dominium model.

There is reason for the CERN theorist's rage other than those mentioned. That laboratory is the one most likely to create stable black-hole material first. Deep under the ground, straddling the border of Switzerland and France the largest machine ever built by human beings nears completion. This machine, LHC, is the cumulative result of many nations, many people, and billions of dollars. The theorist in question is rumored to be one of those potentially up for a Nobel if LHC proves his theories correct. He is deeply vested in its success. To question the basic assumptions of current theories, to him, is tantamount to sacrilege.

LHC is the modern-day Babel. In that parable/historic account, people of many nations forgot their insignificance and the supremacy of God. They put their faith in science and their own prowess. They built the most massive structure ever imagined; only to have it implode on itself and cast them all into oblivion. The parallels could not be stronger. If LHC implodes on itself (by creating stable black-hole material) the mini-trash-compactor synthetic black-hole will begin to feed. It will incorporate adjacent Earth-matter into itself and grow larger. It will become heavy and sink through the layers of the Earth. It will eventually reach the core of the planet, where it will continue to grow. As it feeds, matter that once occupied substantial volume will be compacted down to near nothingness. The Earth will begin to implode upon itself. As the Earth implodes, the crust will sink, rupture, and buckle. The oceans will slosh into areas that sink most quickly and the buckling crust will expose magma, fire, and brimstone. This doomsday cascade would be complete and irreversible once it begins.

Fortunately, at the time these words are written, LHC has not been successful. And there is hope. There is hope that not only will the doomsday scenario be diverted, but also that LHC could be altered to conduct scientific research that does not threaten the existence of all life (current and future) on this planet. In the story of *Horton Hears a Who,* annihilation is averted when all the Who's begin shouting, "We are here, we are here, we are here!!" Essentially, that is what needs to take place today. The author of this work, the readers of this work, and all the rest of humanity are insignificant on their own. But, together we gain significance. If, once this work is read, readers contact their friends and representatives, and if, the readers, their friends, representatives and influential people apply pressure, then activities at CERN, and other laboratories attempting to create black-hole material, can be stopped, reassessed, and the course readjusted. The only way to save ourselves from the impending doom is to choose to act.

## ENDNOTE TO THE INTRODUCTION

This book has been registered with the ISBN agency officially as "science fiction" slash "Science/Nature". Normally, science fiction involves the bending of known understanding. The work does not intend to do that. In fact, this work has been highly researched and cross-referenced with the most current understandings of science reported in the highest esteemed of scientific journals. The "science-fiction" classification was a logistical move. The fact is, there are more readers of science-fiction than there are of works of hard science. This work needs to be read by as many people as possible, as soon as possible.

Besides, it is true that the Dominium model is a hypothesis, which is backed up by known understandings of science. All hypotheses (including Einstein's and currently held CPT Violation) could be considered fictional until they are categorically proven correct. To depict this work with absolute certainty would be to commit the same act of ego-blindness that the Nobel-hungry folks at CERN and Brookhaven have done. Honestly, this author hopes that the conclusions and implications contained within this work can be proven incorrect. If they are correct, the only course of action is to have this work published. Enough people, to achieve critical mass, must be mustered in the short window available, or the End is truly near. For these reasons, the classification of this work as "science-fiction" is the correct move.

## BACKGROUND

Although there is much that current science understands about the Universe, there are many open questions that have no answer under current theories. Among the unanswered are the following questions: What drives the solar wind (a gravity-defying flow of tons of matter per second away from the sun)? How did the giant black hole at the center of the galaxy form? Why didn't the rest of the galaxy collapse also early on in development? If antimatter is so common in high-energy labs, and the Big Bang was the highest energy event ever, why don't we see any evidence for antimatter in our world or observable Universe? Why does the Universe appear to be uniformly distributed? How are the major forces of nature related, or are they related? What is "dark matter" that time after time is measured—between galaxies—where no theory predicted it to be? Although the original goal of this book was only to answer the question concerning the missing antimatter…all these other questions, and more, will eventually be answered in the pages to follow.

The trigger that started these writings occurred at MIT on June 26th 2007. I was in attendance as an MIT professor was giving a lecture entitled "Physics Frontiers." As part of his talk, he made the comment, "Now that we completely understand the behavior of the Higgs' Boson…" Amazed at such a remark, I raised my hand and asked, "How can you say anyone understands the behavior of a particle that no experiment has ever detected or even found traces for its existence?" Shocked that I understood what he was talking about, he backpedaled and stammered. Eventually, he admitted that the Higgs' mathematical models are understood, but evidence for the particle is yet to be detected. After the lecture, I went to talk to the professor about an alternate theory (known mockingly as the "Swiss cheese" model) that offers an alternate explanation of the science of creation immediately after the Big Bang. To my further amazement, he had never heard of this model. Here I was, face-to-face with a PhD professor from MIT, who didn't know as much about the classic theories of his own field as I did. Amazing, because when I learned about this alternate theory at CERN, it was stated that the "Swiss cheese" alternative had been around for over thirty years. This book began as a letter to that professor to teach him about this forgotten model. Personally, I had always favored this solution for the absence of antimatter observations from Earth because it is cleaner and preserves absolute symmetry, unlike the consensus theories to explain the same phenomena. Although "Swiss cheese" was a qualitative model, it always seemed to hold more promise than the, also unverified, consensus theories that are dependent on exotic math and a conflict with the Grand Unifying Theory and Standard Model assumptions for symmetry to exist at all levels.

Once I started writing, very strange things started to happen to me. The letter got longer and longer as I pulled in evidence to verify the "Swiss cheese" hypothesis. Then the letter transformed into a scientific article that I then intended to submit to a journal. Finally, it transformed again into this book that is now before you. Along the way, the process became metaphysical. Visions of connections that I had never before considered popped in front of my eyes as I wrote. At times I felt more like a spectator than a participant in the act of writing. My hands typed what flashed before me…often I couldn't keep up…but I continued on and the scientific connections continued to push the developing Dominium model into territories I had never before considered. Concurrent, and totally illogically, I felt two strong, yet conflicting, welling emotions. On one hand, I felt a great sense of joy, as if this were a new beginning of new age of reconciliation and peace for all mankind. While on the other, there was a huge sense of dread, despair, and looming death.

It wasn't until the transcription of the "Dark Event" did the source of the sense doom hit me: the Dark Event is a period predicted by the Dominium, when all across the Universe small black holes were created. Suddenly, memories of CERN flashed before my eyes: "Small black holes can't exist," I was taught in a CERN lecture. "Only the most massive stars ever become black holes. Therefore, CERN's future plans to generate black hole material inside of LHC are perfectly safe…they will last less than a microsecond before reverting back to common forms of matter." Or so all of those in attendance in the lecture-hall were told. Yet, the model that was being delivered through me was indicating otherwise: small black holes can and do exist, the model asserts. The cascade of visions that followed was so scary that I jumped up out of my chair, smoked a bunch of cigarettes, and paced for a long time. I really didn't want to sit back down, but I forced myself. And this book is the result.

Actually, over the summer two books were written: this one and another, which will only be of use if disaster can be averted. For the time being, we need to focus on the danger at hand. Before that danger can be discussed, the science must be explained. My apologies to those who don't have a background in science, but to explain this model, sometimes the writing is going to get rather technical. To aid in understanding, analogies connected to every-day natural phenomena were used to explain some of the more complex sections. Also, formal logic, in the form of Aristotelian syllogisms, is included to aid the reader if ever you got lost in the narrative. Don't give up.

There will be those who will wish to dub me "delusional" in attempts to overturn this work. Such an ad hominum attack would be fallacious because it would ignore the construct entirely. It is true that I have given credit to divine inspiration, but I am not alone in my experiences. Refer to the quotes on the back cover: Einstein, Sagan, and Curie are all quoted as feeling a divine sense of wonder and beauty as a result of their scientific endeavors. Throughout my education, I have felt closer to God with each new scientific understanding. To me, even the notion of evolution is compatible with the narrative of Genesis (ignoring time, after all what is "time" to an eternal presence?), because the sequencing of events is so similar. Who I am, what I believe, and what I experienced, doesn't matter. The construct is sound; the premises are valid; the science is referenced by the most current sources from reputable journals; and the implications are plausible—and frightening.

Although the reading level is often very high, it is important that the reader attempt to understand. Even more important than the attempt to understand the most technical science in this model, is the call to action. Yes, LHC might create a stabile black hole that would consume the Earth. However, this fate is not a foregone conclusion. Although CERN is the most ivory of towers, they do not operate completely autonomously. This project can be stopped, but it going to take a lot of people to stand up before they will be noticed. No-one should ever think that they are so small that they don't make a difference. True as individuals, we are all pretty insignificant. However, united we can yield more power than all the Nobel laureates in the world. Remember, the world of the Who's was only saved once the smallest of the small joined in and uttered a "Yopp." Once you've read this book: talk to whoever will listen; encourage them to read for themselves; write to your local pubic officials; follow your heart; and be proactive. Good luck and may God be with you.

# OVERVIEW

This book has been divided into several sections to aid in processing this material for the reader. The first section, pages 1-15, establishes the premises (axioms and hypotheses) used in constructing this model. As Einstein is quoted on the back cover, "The grand aim of all science is to cover the greatest number of empirical facts by logical deduction from the smallest number of hypotheses or axioms." Although this model labels nine premises, there are actually only five true fundamental premises in this model (the other four represent either basic conclusions resulting from more basic premises or, in the case of premise 1, a highlight of the current gap in scientific understanding). The model's cornerstone premises are:

a. Premise #2: Assumed supersymmetry between gravitational and electrical systems.
b. Premise #4: All black-holes form as the result of an imbalance of forces.
c. Premise #5: Use of strong fields during experimentation results in masking effects.
d. Premise #6: During the Big Bang, matter and antimatter were created equally.
e. Premise #7: The Universe original matrix was a heterogeneous mix of "imperfections."

From these five fundamental premises, the model is built through logical deduction, as prescribed by Einstein, in pages 16-91. This process is formalized through the use of classic Aristotelian syllogisms. Within this portion of the book, forty-seven separate conclusions are established through this process of deductive reasoning. Together, these conclusions explain and predict almost all "anomalous" phenomena currently unexplainable given previously accepted models.

The appendixes are designed to assist the reader with any questions that might be posed in consideration of the Dominium model. Appendix 1 and 2 show the drama and arguments posted on the Scientific American community blog-space in December 2007 and July 2008. Neither of these first two appendices were edited to preserve their original integrity. Appendix 3 is a collection of asides not necessarily connected to the main deductive flow of the construct but are rebuttals to past comments of "experts" in the field of theoretical physics. Within Appendix 3 rebuttals are presented that have been voiced by past detractors against the notion that matter and antimatter have a gravitationally repulsive relationship, and the flaws of these arguments are exposed. Appendix 4 delivers both a prediction from Kuhn concerning the expected reaction that this model will receive, as well as a quick overview of the type of deductive process employed within the construct. Appendix 5 provides the final conclusions, a Sept 24th 2008 Press Release based on breaking news surrounding the first attempt to start up that machine, and the personal feelings that the author. Finally, there are three reference sections: a Delineation of Construct that lists all of the premises and conclusions of the Dominium model, a Glossary of terms used by the Dominium model in lay terms, and a complete Bibliography of modern scientific journal articles cited.

This book was written over a two-month period during the summer of 2007. It, and its companion book, The Dominium Epiphanies (to be published only if LHC is stopped), can only be attributed to divine intervention. That is not to assume that all words and phrases used are divine, but the visions and revelations that produced the bulk of this work can be explained in no other way. Outlined in this first book is the sequence of events that occurred during the creation of the Universe. Also sequenced are the events that will eventually (an extremely long time from now) lead to the collapse of the entire Universe, and a starting over as another Big Bang. Resulting from this new model are implications that the current direction of modern science is conducted under false premises and could lead to a premature end to our planet in the very near future.

The LHC project at CERN is one of the largest, most expensive, and most multinational endeavors mankind has ever undertaken. The project is heralded as a symbol of peace, cooperation, and scientific advancement. However, despite the hype, this project could succeed in destroying everything.

To stop LHC right now it would take a miracle—perhaps the delivery of this book at this time is such a miracle. By last account, the LHC project was due to start up on its maiden run very soon. So it is a race: who can achieve "critical mass" first. For LHC, critical mass refers to the piling up of matter that results in the first (and last) stabile black-hole material created by man. This might not occur on the first try, but as technicians increase the injection of protons, the chances for "success" rise. For this book, critical mass refers to the number of people who have read it, understand it, and act to warn others, or otherwise act to thwart the start-up of LHC. If LHC achieves critical mass first, it's all over. Money won't mean anything; prestige won't mean anything; and all that we've built won't mean anything.

Experiments ongoing at CERN and other labs are of supreme concern because they operate under several false assumptions. The most troubling of these false assumptions is the idea that to form stable black-hole material one needs vast amounts of matter in one place. The model presented here shows that the formation of black-holes (e.g., MBH) need not require large amounts of mass to be stable. The only requirement is to have enough force directed at one particular spot. And LHC is designed to do exactly that. If this machine succeeds in creating a stable sample of black-hole material, no-one will receive a prize, Nobel or otherwise. A doomsday sequence would be initiated that would not be stoppable, leading to the eventual demise of all life, from microbe to man.

The most chilling aspect of the doomsday sequence predicted by the formation of stable black-hole material is how it parallels apocalyptic descriptions written in both Judeo-Christian and Islamic texts. If created, then the synthetic black-hole would feed on any Earth-matter touching it. Soon after its formation it would sink into the crust of the Earth, feeding as it penetrated. The first noticeable result would be the creation of a new volcano at the CERN facility erupting from the borehole left by the sinking black-hole. After that, earthquakes would be felt all over the globe as the planet readjusts to the loss of volume of material in its interior. As the black-hole continued to feed, some areas of the surface would sink faster than others, causing the oceans to rush in causing death and devastation through floods. As the Earth continues to lose internal volume, the earthquakes would intensify, and eventually the crust would buckle and crack open exposing magma, brimstone, and causing fire. The air would become choked with sulfur dioxide and other noxious gases from the multiple volcanoes and rifts. The process would continue until every last rock, fossil, and human artifact is consumed and crushed into virtual nothingness. To those who wish to find comfort in the idea that the process would be quick and painless, this model does not bring solace. Rather, the process would be expected to last a number of years (of terror), because a black-hole at the Earth's center would act like an hour-glass—material would continuously and assuredly fall in, but not be free to fall in all at once. When the process is complete, the only evidence of humankind will be the satellites that will continue to orbit the space where the Earth once was. All this parallels the stories of the apocalypse as told by the Bible and Quran: flood, fire, brimstone, and being sucked down to Hell.

With the publication of this book, the onus of responsibility transfers from the author to the reader. What will you do?

## Part 1: The premises/assumptions used to build the Dominium model

***Premise 1: No understanding exists to describe the gravitational interactions between matter and antimatter.***

**Simply put:** No one knows the true nature of gravity between matter and antimatter. All commonly held theories up until this point follow the assumption that these two materials attract each other.

I once made the mistake of answering a question from my $10^{th}$ grade English teacher by saying, "I assume…" She cut me off mid-sentence and wrote the word "assume" on the board. "Assume!" Mrs. Gross exclaimed. "Assume? How dare you assume anything in my class!" Horrified and speechless, we all watched as she took her chalk and continued on. "Whenever you *assume*," she pronounced, "you make an *ass*" (underlining the first three letters of the word), "out of *you*" (circling the middle "u" of the word), "and *me*!!" (underlining the last two letters of the word). Although I would never dream of crucifying a student in that way, Mrs. Gross was right. Making assumptions is a dangerous affair, especially in science. The earliest theorists made an initial assumption that matter and anti-matter attract. Assumptions *do* cause problems, especially when they're incorrect.

Of course, in the pursuit of science, there is a Catch-22 when it comes to assumptions. Sometimes it is impossible to proceed without laying down what is assumed to be true and what isn't. Einstein's work would not have been possible had he not made some basic assumptions. Confounding the problem is the fact that many of Einstein's conclusions have been proven correct, especially the notion that matter can be converted into energy and vice versa. However, just because some of his conclusions have been proven correct does not mean that all of his assumptions are valid. Although Einstein himself may have been aware which parts of his model were assumptions, he has gained the stature of being unquestionable and absolute in his correctness. The true fault lies not with Einstein himself, but with those who followed. Subsequent theories and theorists have forgotten that one premise especially—that antimatter and matter attract each other—is an assumption rather than a proven fact.

Like all scientific theories, the Dominium model must make some assumptions in order to begin. Because the question of gravitational interaction between matter and antimatter is an unanswered and open question, a model can be developed using either assumption as a basis. The new model takes the opposite and equally reasonable assumption as its hypothesis: matter vs antimatter gravitationally repel.

**Scientific reasoning:** Although this premise will never be "referenced" throughout the body of this work, it is the opening in the gap through which all subsequent scientific discourse is squeezed. This one gap in verifiable understanding is crucial. It will be shown that all current models make the assumption that all gravitational interaction between particles is always one of attraction. If this assumption is incorrect, the common conclusion (that antimatter is attracted to matter) is without base. The central hypothesis of this work is that the assumption is indeed incorrect: matter and antimatter repel one another.

In order for the central hypothesis to be made plausible, this first premise needs to be firmly established. In situations where no definitive understanding exists, one hypothesis is as good as any other. Einstein and friends considered the possibilities of creation where matter and antimatter gravitationally attract; the Dominium model takes a path not taken before, the possibility that matter and antimatter repel. The explanation of this scientific gap in conclusive understanding is given in the following syllogism:

1. All experiments involving the manipulation of antimatter involve the use of either strong electric or magnetic fields, which are many orders of magnitude above the expected effects of gravity. Rest gravitational attraction is $10^{43}$ times weaker than electric repulsion forces between electrons and $10^{36}$ times weaker than that between protons. Thus, the mechanically derived fields used in experiments are many times stronger than rest forces.
2. So far, all particle physics experiments have been conducted on Earth, which is awash in matter (high particle density); therefore, annihilation quickly consumes the samples of antimatter produced, unless they are held within a vacuum.

• Therefore, one cannot surmise what, if any, gravitational interactions matter and antimatter might have with one another.

(And in fact, those who work with antimatter accept that we don't currently know what, if any, gravitational interactions antimatter might have with matter.[Holscheiter])

The preceding syllogism is based on inherent design difficulties in past experiments involving antihydrogen. Even in cases where antihydrogen, antiprotons, and positrons have been captured and cooled, the storage structures, penning traps, must use strong magnetic fields to hold the antimatter in stasis.[Bromley] Because the resulting confinement forces are strong enough to decelerate and contain antihydrogen in one desired location,[Kellerbauer] it is not an unsafe assumption to say that the comparatively weak effect of gravity would be masked in these tests.[Poth]

***Premise 2: All phenomena that occur in electrical systems are directly applicable as a predictor of gravitational behavior, if the conditions are tightly analogous, because of supersymmetry.***

**Simply put:** Gravitational and electrical systems will behave similarly. There is a theory, which was dear to Einstein's heart though he never managed to prove it, called the "Grand Unifying Theory." Within this theory the major forces of nature (electric, magnetic, gravitational, nuclear, etc.) are tied together in a predictable and neat manner. On Earth, the scale of scientific experimentation lends itself to the study of electrical systems. We have learned a great deal about how electrical systems work, behave, and interact. Our knowledge is the basis of the disciplines of physical chemistry, nanotechnology, and microprocessing. However, the scale of gravitational systems is much too large for humans to experiment on or manip-pulate. Rather, we observe from observatories phenomena that occur naturally. As a result, we do not have a working basis of how gravitational systems interact with one another. One assumption of the Dominium model is that on the basis of the Grand Unifying Theory, gravitational and electrical systems will possess simil-larities. These similarities will lead to behaviors, interactions, and outcomes in gravitational systems that are similar to those that have been measured in electrical systems.

**Scientific reasoning:** One of the goals of this model is to show multiple examples of supersymmetry between gravitational and electric forces. However, in this very initial stage, supersymmetry is put forward as a founding premise. This might seem to be a bit presumptuous, but it's not. If there is an inherent connection between these two forces then that connection will appear repeatedly, from beginning to end. No model has ever spanned the time from Big Bang to Big Bang . . . and there are points where the uncertainty of the next step is blinding. Premise 2 will be the most crucial of all the premises, because with it predictions can be categorically deduced projecting what the system will do next. Without this tool, this model could never have been built. Once built, the construct itself reveals how obvious the assumption of Premise 2 is.

Science exploration and discovery is often described with the metaphor of an elephant hidden in a big box. Each scientist (lab) touches a different part of the animal, without seeing it, and comes to a different conclusion. This explains why the process can become stymied. However, if a scientist were to "cheat" by receiving word before conducting a test that it was actually an elephant that he/she was about to probe, then finding evidence to that effect would be easy. That is what the Dominium model is doing by establishing Premise 2. If however, the scientist tried to "cheat" and received word that inside the box there was a camel (though it was in fact an elephant), he'd run his tests and maybe even pronounce the victory of discovering a camel, even while tests from other labs would show anomalies to the camel hypothesis. This scenario is analogous to what has happened with CPT Violation theories (discussed below), which, although they are accepted by the scientific community, are at a loss to describe multiple anomalies. The adoption of Premise 2 by the Dominium model will be vindicated by its seamless predictions of all events, including current anomalies, from Big Bang to Big Bang.

Supersymmetry between these two forces, electromagnetic and gravitational, has been suspected for a long while. Anecdotally, we can note that first-year physics students recognize the similarities between Newton's Universal Law of Gravita-tion and Coulomb's Law. Both laws describe field forces: forces that can be felt at a distance regardless of the medium (or lack of) between the two "things" in question. Not only are these two forces the first examples of field forces presented to first-year students, but also the equations used to describe the two are extremely similar in form:

$$F = \frac{kq_1q_2}{d^2}$$

Coulomb's Law

$$F = \frac{Gm_1m_2}{d^2}$$

Newton's Universal Law of Gravitation

Both show the relationship two "things" feel at a distance from one another. One might point to the "G" versus the "k" in the two equations as being a significant difference. However, closer inspection reveals that both "G" and "k" are constants, which effectively balance out the effects caused by the unavoidable arbitrary nature of units used by humans. Even the metric system is an arbitrary system. There is a theory (the Grand Unifying Theory) that all forces in nature are bound together in some sort of explainable fashion. Ultimately an equation could be drawn up which explains this relationship. However, the "true" unit of, say, length would be the value that most correctly simplifies this overall equation, taking into account the relative relationships/values with all other fundamental aspects of nature, e.g., charge, mass, time, etc. Since we do not/cannot know these relationships before beginning any analysis, we've settled on using agreed upon standardized values based on some reproducible aspect of nature. A net result of these arbitrary units is the need to utilize "constants" to turn a proportional relationship into an equality.

Note that fundamentally, the relationships shown in these two equations are extremely close: they are direct relationships between the force felt and the magnitude of each of the "things" analyzed, and an inverse relationship between the force felt and the distance squared. In other words, the larger the "thing" the more force felt, while the smaller it is the less force felt. And also, if the distance were to triple, force would reduce by a factor of nine; if the distance is cut in half the force will quadruple. Equations that describe completely different phenomena yet contain the same fundamental framework are extremely rare. This rarity provides our first hint at supersymmetry. Therefore, Premise 2 will be accepted as valid for the purpose of building the Dominium model. With this premise comes the idea that similarities between these two forces extend well beyond surface appearances of the equations used to define them. Therefore, examples derived of how electrically charged particles behave under a given set of criteria will be accepted as a valid predictor of how gravitationally driven particles would be driven under identically analogous conditions.

The conclusion in the preceding paragraph is the first step in describing supersymmetry between gravitational and electrical forces. By the end of this book, supersymmetry between gravitational and electric forces will be defined from the initial

Big Bang to subsequent Big Bang. However, for the time being, the model will use this premise by viewing examples of electrical phenomena as a predictor of the behavior of gravitationally driven systems, to explore the ramifications and to see where such repercussions lead. Proof of the validity of Premise 2 will come in the form of a complete model that spans from Big Bang to Big Bang and in the prediction for and explanations of phenomena that are currently without explanation given current understandings. Along the way, if Premise 2 is utilized, then a subsequent conclusion will be drawn that the two phenomena being described are supersymmetric equivalents of each other.

***Premise 3: Gravitational forces will experience exponential increases in effect at very close ranges.***

**Simply put:** Just as has been measured in electrical systems, at very close ranges the particles responsible for gravity will experience a "ballooning effect."

**Scientific reasoning:** Laboratory tests studying the effects on force of two charges at very close ranges show that Coulomb's Law relationships no longer hold true to the classic equation in these conditions. Specifically, the relationship between force felt and the distance between particles changes. If two similarly charged particles are moved increasingly closer together, then the repulsion force balloons at a rate much higher than expected. This ballooning effect has been measured at higher than $F \propto d^{-7}$. Although this has been measured in a lab, it is not conclusive where this ballooning effect will end if even more minute distances were to be tested. The syllogism for Premise 3 is:

a. Electric repulsion forces are shown to balloon at extremely close ranges.
b. Premise 2: All phenomena that occur in electrical systems are directly applicable as a predictor of gravitational behavior.
c. Therefore: Gravitational forces will increase exponentially (balloon) at very close ranges. (Premise 3)

***Premise 4: All black-holes form as the result of an imbalance of forces: Gravitational effects favor collapse, while other forces act to prevent collapse.***

**Simply put:** The formation of a black-hole is the result of an imbalance, just like the children's games: Jenga, Don't Break the Ice, or Don't Spill the Beans. All three of these toys employ the same principle: a system begins in a stable position. By playing the game, the system becomes increasingly unstable. The game concludes when a tipping point has been reached and the entire system crashes in upon itself. The analogy works well to explain how stars go from stable systems, to supernova, to black-hole. The star begins as a stable system. The game being played is *fusion*, which makes the star increasingly more dense. As the game goes on, a tipping point is eventually reached where the entire system crashes in upon itself. Just as in the children's games, once the crash has occurred, it cannot be undone in any straightforwardly natural fashion.

**Scientific reasoning:** True, concerning the creation of black-holes, there are many other models. Some models are very complex, for example those that define the dynamics of a black-hole through multiple dimensions and complex mathematical constructs.(Aspinwall, Gergely, etc) Applying Occam's razor, the Dominium model is going to use the simplest possible definition of the process of creating a black-hole. For the sake of this analysis the simplest scenario—an imbalance of forces—for the creation of a black-hole will be used. The most typical example of a black-hole is that of one formed from a large star, described as follows:

1. In any big star, many times bigger than our own, fusion has been going on for eons.
2. Fusion is the process of taking small-sized atoms, *fusing* them, and making bigger ones.
3. Assume the general assumption for Gas Laws holds for solar plasmas: all atoms take up the same amount of space, regardless of size/density.
4. Therefore, the big stars are getting denser.
5. As the star becomes denser, the gravitational pressure increases.
6. Electrostatic pressure also increases.
7. The rate of increase in gravitational pressure is faster than electrostatic pressure.
8. As this imbalance in rate of change continues, gravitational forces exceed electrostatic.
9. Eventually, supersaturation of imbalance occurs and the system collapses in upon itself.

With the basic schedule of events listed above, to create a black-hole one need only create an imbalance of forces. In the above list of the normal sequence leading to the production of a black-hole, the force listed that ultimately causes the formation of a black-hole is that of gravitational attraction. Let it be noted that the description given is for a matter-based black-hole formed at the end of the life span of a large star such as might be found in the Milky Way. Let it also be noted that the above sequence does not preclude the formation of a black-hole through other scenarios of imbalanced forces. It is simply a sequence of steps that many believe to be responsible for the formation of a black-hole. Because it is the simplest theory to account for this phenomenon, it will be used as a premise for future situations to be considered.

***Premise 5: Use of strong electric and/or magnetic fields during experimentation results in a masking effect.***

**Simply put:** The machines used in current antimatter experiments produce field forces that would have a much stronger effect on the particles being studied than natural gravitational repulsion between antimatter and the Earth.

**Scientific reasoning:** Although this concept has already been introduced and discussed in this paper, it is such an important point that it needs to be reiterated. Fields used to accelerate, contain, and manipulate particles (with special notice to antimatter) are all going to impart force contributions many times greater than forces felt by gravity. The effects of gravity are expected to be less by $2.3 \times 10^{39}$:

23,000,000,000,000,000,000,000,000,000,000,000,000,000[Miller]

times smaller. Even with the most sensitive detectors available today, the relatively and vastly minute of gravity compared to other forces present could lead to masked effects.[Poth]

***Premise 6: During the Big Bang, energy converted into matter and antimatter symmetrically, i.e., in equal amounts.***

**Simply put:** The Big Bang would be expected to function like all high-energy experiments done to date. Every time that $E=mc^2$ (energy transformed into mass) has been observed and documented, the event produces exact and equal amounts of both matter and antimatter. This well-documented outcome is referred to as "pair production." In high-energy physics laboratories across the globe, pair production is a normal, predictable, and usual outcome of a high-energy event. There is no reason to believe that the Big Bang would be any exception.

**Scientific reasoning:** Symmetry is assumed beyond all else to be true for the Big Bang. This is a primary premise because in high-energy experiments, pair production is the standard occurrence and manifestation of Einstein's famous equation, $E=mc^2$. When energy is converted into matter it is done so symmetrically. If a positively charged particle is formed, a negative one will also be formed. In this standard occurrence of high-energy experiments, the scenario is always the same: equivalent charges, equivalent masses, and one the antiparticle of the other. The true task of particle physicists studying collisions is to try to figure out what, if any, recorded events *are not* tied to a pair production event . . . it's that common and predictable.

In very high-energy experiments exotic particles are often produced that give off more energy than is stable in the conditions of the lab. These particles will precipitate a cascade of events, going through pair production events as part of the decay, until only particles that are stable under laboratory conditions remain, e.g., electrons, protons, neutrinos, and photons. There does exist a rare and disputed cascade event that does not seem to be symmetric, but it is a decay event and not a production one. This issue will be discussed later. The process of converting energy into matter seems always to occur symmetrically: equal amounts of matter and antimatter are *always* produced.

**Introducing the central dilemma—How to explain the fate of the missing amounts of antimatter visible in our modern Universe**

**Simply put:** Einstein's famous equation also works in reverse: $mc^2=E$. This reverse-direction process is called an "annihilation event." Mass can be converted into energy if a piece of antimatter and a piece of matter collide. If equal amounts of matter and antimatter were produced during the Big Bang, why didn't all of creation eventually convert back to energy via annihilation events? The dilemma is caused by the "traditional assumption" that gravity is always an attraction force.

**Scientific reasoning:** Assuming that equal amounts of matter/antimatter, charge, and other things of the Universe were created in equal portions as dictated by symmetry, a small puzzle arises: what happened to the antimatter? No antimatter (other than the abundance of positrons in stars like our Sun) is observable naturally in our surroundings. If it was created in equal amounts, why aren't stray antiparticles whizzing around with our observable matter? Why don't we see a history of annihilation events in the sky? There are two main models to explain this.

**CPT (C-charge inversion; P-parity inversion; T-time reversal) Symmetry Violation Theory—the "accepted" theory**

**Simply put:** The commonly held "solution" to the absence of antimatter dilemma is a theory that states that the missing antimatter essentially "disappeared." Believers in this theory use much higher language and justifications than that, but that is what their arguments boil down to. The suggested pathway of this "disappearance" is not through the accepted $mc^2=E$ route, either; rather it involves extremely complicated, convoluted, and exotic math.

The Dominium model does not accept these currently held "solutions." For one, the explanation and the mathematical models are extremely complex and convoluted. As the early philosopher Occam pointed out, the more complex and convoluted an explanation, the less likely it is to be correct. Also, these types of exotic decays are not well documented. In comparison to observed pair production events, only a minute thimbleful of such events have ever been recorded. Those "verifications" that have been recorded could be attributed to measurement or machine error. Even if the "verification" events have been recorded correctly, there is still the possibility of a symmetric anti-event that would maintain the balance and symmetry predicted by the basic idea behind the Grand Unifying Theory. However, because CPT Violation is the accepted "solution" to this dilemma, it will be discussed further before being put aside and replaced with the Dominium model.

**Scientific Reasoning:** The mathematical construct of CPT Violation, first introduced in 1967 by Andrei Sakharov,[PPRAC Antimatter] relies on possible differential behavior (parity violation) of the decay of matter versus antimatter mesons, leading, very early on, to an entirely asymmetric Universe consisting only of matter. This theory accepts *Premise 4: During the Big Bang, energy converted into matter and antimatter symmetrically, i.e., in equal amounts*; however, it states that what happened next is the reason why we see an entirely mass-based universe.

According to this theory, once energy was converted into masses of both matter and antimatter, the "exotic" mesons (now formed in the lab) would be the norm in nature rather than the outlier. When these particles decayed, pathways that led to asymmetric decay caused the asymmetric matter-only Universe that we see today.

Supporters of CPT Violation's assessment of an asymmetric matter-only based Universe will point to experiments conducted at the BaBar project at SLAC, which aim to catalog asymmetric decay of B-mesons[(PPARC-- Mystery...)] The asymmetric decay of B-mesons has been observed in some experimental results, although it is not a common observation. Some believe that the rare occurrences "observed" are actually misread/faulty data, the result of "noise" within the data, a single faulty detector, or some other malfunction of the machinery. Once machine error is ruled out, the other problem with using this data as a foundational premise for CPT Violation is that the scope of the experiment is not wide enough. Perhaps there is some other, yet to be identified type of decay asymmetrically favoring antimatter over matter, thereby balancing out B-meson asymmetry and preserving overall symmetry throughout the Universe.

***Premise 7: The Universe's original matrix contained imperfections in the random heterogeneous matrix.***

**Simply put:** Although the known process $E=mc^2$ (energy transformed into mass) produces quantities of matter and antimatter in precise and equal amounts, there is no reason to believe that the distribution of this material would be even and ordered. Rather, statistical variations in the concentration of these two types of mass would be expected.

**Scientific reasoning:** There does exist another theory of the Big Bang that does not rely on CPT Violation to explain the absence of antimatter in this quadrant of the Universe: the Swiss-cheese model. This theory is based on Alan Guth's "inflationary model" that suggests that "imperfections" might have spread through space as matter began to appear where previously there was energy.[PPARCBIGBANG] This heterogeneity in the distribution of matter became the seeds that formed the galaxies that we see today. Although Mr. Guth himself favors the asymmetric solution to the dilemma of the missing matter,[Guth] the Swiss-cheese model follows his ideas, yet maintains symmetry. Unfortunately, this particular theory is discounted by the scientific community because it is qualitative in nature.

The Swiss-cheese model adds a twist to Mr. Guth's model: some of the "imperfections" would be matter-dominated areas, while others would be antimatter based. According to this model: although creation led to equal amounts of antimatter and matter, distribution was slightly and randomly heterogeneous, and annihilations occurred very quickly early on and then abruptly stopped as areas became cleared of either of the two types. These areas of clustering would resemble the holes of a giant block of Swiss cheese, while the "cheese" would essentially be nothingness. These holes then matured into the various galaxies; therefore some galaxies would be exclusively composed of matter and some exclusively of antimatter. Light coming from an antimatter star would be indistinguishable from that emitted from a matter star because light is the antiparticle of itself. Unfortunately, the original authors of the Swiss-cheese model stopped at this point. Their theory could not explain any further. Because it was a qualitative analysis, it was derided for its lack of mathematical foundations. Because it could not make any further predictions, it was labeled as an unsubstantiated dead-end.

The moniker "Swiss cheese" was also unfortunate for the survival of this theory within the scientific community's consciousness. A theory referred to in such a way was easy to make fun of—after all, this theory was "full of holes." On closer inspection, this model holds a great deal of promise. After all, it stands squarely with all Standard Model assumptions, including CPT *symmetry*.

This theory has been around for a long time. Several labs have been working on a project that could easily disprove or prove the Swiss-cheese model: CERN's antihydrogen project and LEAR. A goal of this lab is to produce antihydrogen and other forms of antimatter to study their dynamics in relation to fundamental questions.[Abdullah, Bromley, Holzcheiter, Poth] The reason why forming antihydrogen could prove the Swiss cheese model is as follows: if one can create antihydrogen, and one can produce the spectral signature for antihydrogen, then one can compare it to that of known matter-hydrogen. CPT *symmetry*, as prescribed by the Dominium model, is one of the pillars of the Standard Model of elementary particle physics and requires hydrogen and antihydrogen to have identical spectral lines.[Waltz] Unfortunately, that particular test has not yet been successfully accomplished.[Holzcheiter] If the two spectral signatures match up exactly, the Swiss-cheese model is proven; otherwise if the spectral signature of antihydrogen is significantly different from that of common hydrogen, then the Swiss-cheese model is proven incorrect. The reasoning is that it's relatively simple to determine the composition of the star by using the spectral bands of the light produced and then compare this to the spectral fingerprints of known elements. Through this type of observation, the composition of many stars and galaxies has been determined. Conclusions show "hydrogen" to be the dominant component; this observation has been repeated and its conclusion is an accepted fact within the scientific community. However, if antihydrogen produces exactly the same spectral signature as common hydrogen, and if symmetry is exact in all forms (e.g. antifusion occurring in antimatter stars), then observed galaxies could indeed be made of antimatter. If the match were not exact, then there are no antimatter stars undergoing antifusion, and antihydrogen wouldn't be expected to exist anywhere in the visible Universe, because observatories achieve consistent readings of "hydrogen's" spectral signature and no anomalies wherever they point their telescopes.

Although Walter Oelert and his lab are credited as having synthesized antihydrogen in 1995,[PPARC AntimatterMHS] what they produced was only around long enough for them to detect its presence. In early experiments a problem arose once antihydrogen was created: it quickly becomes electrically neutral, and therefore the electric containment fields no longer have an effect. Any kinetic energy (and all the atoms formed were still quite energetic) would cause the new sample to crash into a matter-dense wall of the instrument being used and annihilate themselves. This problem has since been resolved through the use of strong magnetic fields to stop the antihydrogen ions in one vacuum-protected spot in a containment device known as a "penning trap." [(Abdullah, Bromley, Holzcheiter, Poth)] Again, the use of strong fields, this time magnetic, has been identified as a possible mask of any/all minute effects of gravity felt by the antiatom.[Poth]

Attempts at spectral analysis have yet to be successful.[(Abdullah, Bromley, Holzcheiter, Poth)] Common spectral analysis is a very basic lab practiced in high schools across the globe. The only things required are a tube filled with low-density gas and a high-voltage source. The high voltage will cause electrons to hop from gas molecule to gas molecule, bringing them to an excited state, and then when the excited electron falls back down to a more stable energy level it will emit a photon with an exact and set value of the energy it absorbed. Comparing the spectral lines to a standardized key allows students to be able to positively discern what material is inside each tube being studied.

Conducting the same test with antimatter is much more difficult. The low-density part is not a problem; the next part is hard. How does one set up a voltage where positrons get pulled across by a field, as electrons are pulled across in the common spectral experiment? We know how to shoot a beam of positrons. Although particle beams are often described in terms of currents (charge per second), they are really currents based on inertia, not electric potential. If an accelerated beam of positrons were shot into a penning trap containing antiprotons, the results would be meaningless. In such a case, the two particles would be expected to collide and undergo pair production events as energy is converted into mass, $E=mc^2$. A pair production event is nowhere near comparison to a spectral signature and such data would be inapplicable. The missing required element is a voltage based on antimatter where positrons feel the pull of electric potential and then proceed to hop across the antihydrogens, causing them to rise to excited states, and then ultimately fall back down to ground-state and reveal their characteristic fingerprint.

Regardless of the current lack of information regarding the exact spectral signature of antihydrogen, antihydrogen has been made. Researchers are perfecting their methodology in the production, storage, and cooling of antihydrogen.[(Abdullah, Bromley, Holzcheiter, Poth)] These facts lend support to the foundations of the Dominium model. This model requires symmetry at all levels. The CPT Violation hypothesis holds that asymmetry between properties of matter versus antimatter accounts for a stable matter-only modern Universe. Now, consider the results from the lab: antihydrogen *can be/is* created in the lab, and it *is* held stable in vacuum-protected penning traps for long periods of time. If it had been proven that antihydrogen could not be synthesized in the lab, or that it was extremely unstable, models favoring asymmetry based on CPT Violation assumptions would be supported. The fact that antihydrogen *can* be produced supports models based on the fundamental idea of symmetry as a strong determinant of the existing world. The Dominium model is in total agreement with the Standard Model's foundation of CPT symmetry, and is therefore supported by laboratory success in creation and long-term storage of antihydrogen.

Detractors of the Swiss-cheese model point to the absence of annihilation events in the sky's history as proof of the model's invalidity. Such reasoning states that the fireball would have created all sorts of trajectories for both matter and antimatter; and therefore residuals that did not immediately become consumed (for surely there would have been residuals) would create large random annihilation events that would be visible from Earth's observatories. However, searching the sky reveals no events like those that would be expected of a massive annihilation of a random large matter object striking a random large antimatter object. As a syllogism, their argument appears as the following:

a. Suppose all mass (matter and antimatter) lasted in equal quantities during the Big Bang as proposed by the "Swiss cheese" model.
b. Suppose all particles produced would have high and random trajectories.
c. Suppose equal amounts of matter and antimatter remained after the Big Bang.
d. Then, some mass of for both matter and antimatter would have been in close proximity and would have high and random trajectories.
e. All matter and antimatter annihilate in the laboratory.
f. Therefore, owing to randomness, there should be a history of random annihilations in nature.
g. No history of annihilation events is recorded.
h. Therefore, premise "a" is incorrect, and all mass (matter and antimatter) *did not* last in equal quantities during creation.
i. Therefore, the "Swiss cheese" argument "is full of holes." (A physicists' joke)

Although this construct appears to be completely sound, on closer inspection there is a flaw. Premise "d" is too limited in scope. Laboratory conditions do not represent all possible scenarios in which antimatter and matter might coexist. The condition that is of supreme importance concerns ratios of matter vs. antimatter locally present at the time of the test. This is the most crucial condition to understand, because expected conditions of the early Universe would be 50% matter and 50% antimatter. However, in the matter-rich environment of the Earth, ratios of antimatter created versus matter within the immediate vicinity are extremely skewed. For example, if a lab were to be able to create and store $10^{13}$ particles of antimatter, within a meter of the storage position of these particles there could easily be well over $10^{27}$ matter-based particles. This means that the ratio of antimatter to matter that is locally present is $1:10^{14}$—in other words one to one hundred trillion—and results from such tests would not be a good predictor of systems closer to 1:1.

Another failing of current/past laboratory experiments is that they are often limited in scope, with respect to the phase/type of particles present. Current exploration almost always involves a diffuse "gas" of antiparticles, in a penning trap for example, surrounded by a "solid" matter machine. Gas-gas interactions/conditions would be most beneficial to our understanding and observations, because the superheated plasma of the early fireball, though unique unto itself, would be expected to behave more like a gas than a solid. Therefore, it is safe to conclude that laboratory findings are limited in scope concerning the diversity of conditions under which matter and antimatter could coexist at close proximity. Therefore, premise "d" is not true. But if we remove premise "d," the detractor syllogism cannot be continued on to its conclusions.

**Part 2 of the Dominium model:**
**Predictions and explanations of events that occurred from the Big Bang to the formation of the modern Universe**

Paradigm challenge: There is a saying that many, if not most, chemistry teachers offer their classes at least once in their careers: "Opposites attract." This dogma is so ingrained in our society that it is even quoted by social conservatives to justify edicts. What if this is true of electric charges, but not true of all central forces?

Questioning this paradigm opens up new solutions. For instance, what if "Like attracts like; and opposites repel" is a valid scenario? This would be an opening for an entirely new understanding of gravity. The current dogma is that gravity is always "an attractive force"—but what if that's only part of the truth?

This shall be the only point in the entirety of this book that a conjecture is presented that cannot be backed up with either reference to more primary premises or with an analogy from nature showing plausibility. The following is a logical construct based on that one hypothetical opening of a single paradigm.

## Hypothesis:
## Gravitational-opposites (matter/antimatter) repel one another, while gravitational-likes attract one another.

***Conclusion 1: The time at which gravitational repulsion forces were the strongest was during the initial moments just after the Big Bang.***

What if opposites (matter and antimatter) repel one another? One can only conclude that the highest impact of this relationship would be felt during the first moments of the Big Bang—when distances between "everything" were very short. Consider the following syllogism:

a. Suppose gravitational-opposites (matter and antimatter) repel.
b. Gravitational forces will experience exponential increases in effect at very close ranges. (Premise 3)
c. Immediately after the Big Bang was the time at which everything was closest together.
d. Therefore, the time at which gravitational repulsion forces were the strongest was during the initial moments right after the Big Bang.

***Conclusion 2: Immiscibility: Opposites-repulsion force would lead to a state of immiscibility below a minimum energy.***

If the embryonic galaxies did repel each other, then, at any given moment, they would be heading away from their nearest opposite. While accelerating away from the first opposite, if an embryonic galaxy came across a second opposite there would be a dynamic, repulsion force, which would act to prevent collisions. However, it is also conceivable that if energies were high enough, gravitational repulsion forces could be overcome. If energy is a determinant in this process, then it follows that below a certain minimum energy gravitational repulsion forces would act to prevent collisions between neighbors. When neighbors stopped colliding, they would be said to be "immiscible," in other words, "not mixable." Collisions represent mixing; without the collisions, there would be no more mixing. Webster's lists "immiscible" as: "incapable of mixing or attaining homogeneity." This word describes perfectly what happened to the Big Bang fireball below a certain energy-density threshold.

a. Hypothesis: Gravitational-opposites (matter and antimatter) repel, while gravitational-likes attract one another.
b. Sufficiently high energies could overcome any level of repulsion force.
c. Then below a minimum energy, repulsion tendencies will be stronger than attractive ones, keeping opposites immiscible.

Once the combination of total regional energy density drops below a critical level, boundaries would be defined and the "Period of Immiscibility" would begin.

Hence gravitational immiscibility would be an energy-dependent function. Energy-dependent functions are found throughout nature—within chemistry, biology, and physics. Such analogues add a degree of plausibility to the above conclusion; the fact that they are numerous adds further support. And the fact that many energy-dependent functions are electrically driven provides initial evidence for supersymmetry between electrically and gravitationally driven systems.

***Model-specific conclusion 2a: Kinetic energy will be the standard form of energy used by this model in connection to conditions for immiscibility vs. miscibility.***

Although there are many forms of energy, for the sake of this model, KE (kinetic energy) will be viewed as a fundamentally important aspect of immiscibility. Reason 1: It is the most basic form of energy already used to describe the behavior of objects that gravitationally interact. Reason 2: KE was already used in the description of the conditions of the early Big Bang in Conclusion 2: collisions assume velocities. Newtonian physics is saturated with applications and examples where gravitational interactions intertwine with kinetic energies within a given system; it is, therefore, acceptable for this model to make that connection. For example, gravitational acceleration is what often occurs with gravitational interactions (consider Newton's famous apple). The greater the speed, the greater the kinetic energy. In terms of energy relationships: "The gravitational potential energy was converted into kinetic energy in a conserved manner." Examples like this are at the very heart of Newtonian physics.

Another reason for assuming that kinetic energy is an important determinant of immiscibility is that with this assumption, most annihilation events in the lab can be explained. Annihilation events are cases when boundaries between matter and antimatter don't hold (i.e., they became miscible). Therefore, one goal of this model is to identify which aspects of past laboratory test design led to conditions of miscibility, and therefore annihilations, within the lab.

A binding premise, not stated earlier, is that predictions of this model need to correlate to both experimental and observational data. One factor extremely common to most past experiments involving antimatter is that the antimatter produced was the result of pair production in high-energy collisions. Because the antimatter was produced in this manner, and because momentum is conserved, resulting particles had high energy. We also know for such tests that antimatter produced is short lived if a vacuum does not protect it. It is safe to conclude that the large values of kinetic energy contributed to the annihilation event. The effects of KE would be strong, if the KE level was high. The stronger the KE, the greater the masking of any/all gravitational interactions, which could then lead to evidence being buried. Either way, annihilations in the lab would not count as evidence of antimatter and matter "seeking each other out," nor would it support the Dominium assertion that matter and antimatter avoid one another. Essentially, annihilations in the lab are of trivial concern to this model, because they are caused by forces and situations not associated with gravitational interplay.

***Model-specific Conclusion 2b: Particle density of the medium being traveled through will be a standard used by this model in connection with conditions for immiscibility/miscibility.***

Although high kinetic energy in laboratory-generated antimatter accounts for why most annihilations occur, it does not account for annihilations occurring under conditions of low-energy levels, such as those currently being conducted at facilities like LEAR at CERN. Assuming model-specific Conclusion 1a is correct, the fact that annihilations also occur at low energies implies that another factor is a determinant for the line dividing miscible and immiscible conditions. The most likely other factor is the particle density of the material to be passed through. Penning traps are created within large machines, which maintain an ultra-high vacuum and the fields that contain the antiparticles. These machines consist of composites of solid matter-type materials—metals, glass, plastics, etc. Solids are highly dense and tightly bound at the atomic level. If a wayward antiparticle escaped out of the magnetic containment of the penning trap, eventually it would have to attempt to pass through a solid (given net forces acting on it) if it were to escape. However, no particles with mass can pass through a solid easily without interaction. To be able to squeeze through without interaction would be what one expects under immiscible conditions. Because an interaction occurs, the antimatter is annihilated, and therefore the density of the machine lowers the kinetic energy (KE) requirement for miscibility.

Thus the conclusion that antimatter and matter "seek each other out," as I've heard lab-techs say, is a false conclusion not supported by evidence. Whether antimatter is attracted to matter or repelled by it, annihilation would be a certain and swift outcome.

***Conclusion 2c: The phenomena of immiscibility and phase change between gas and plasma are supersymmetric equivalents.***

The advent of immiscibility is vital to this model. Immiscibility is essentially a set of conditions below which the dynamic changes and there is a sorting of type. This is such a profoundly important point that if supersymmetry does exist, it would be expected that there would be some sort of electrically driven equivalent phenomenon. This importance is defined by Premise 2: *"All phenomena that occur in electrical systems are directly applicable as a predictor of gravitational behavior, if the conditions are tightly analogous."* The phenomenon that seems to match is the phase change between the gas and plasma states, i.e., coupled and uncoupled electrical states of matter. The coupled state is the more familiar to most people: electrons are held in orbit around a central nucleus. In the uncoupled state, both electrons and protons exist totally free of the other—there is no linkage. Another way to view this phenomenon is that the uncoupled state is higher KE, thoroughly disordered, and heterogeneous, while the coupled state is of a more ordered design. Equivalently, in the Dominium model, the condition of miscibility is a state of higher KE, thoroughly disordered, and heterogeneous, while the immiscibility state is a more ordered one. Part of both phenomena is a sorting: charge-type in the case of coupling, and matter-type in the case of immiscibility. Therefore, because of the closeness in parallels between these two phenomena, Premise 2 can safely be employed and the identification of supersymmetric equivalency can be inferred.

***Conclusion 3: Once conditions for immiscibility have been met, like particles will coalesce.***

Once immiscibility occurs, annihilations will cease. However, interactions between neighboring grains will continue. Like grains will be allowed to collide and consume like grains, but not with opposites. Other models employ similar methods to describe the formation of galaxies as being built over time from independent fragments.[Gao] True, we have photographs of galaxies colliding,[NASA #02094] but with this paradigm-shift, they'd be expected to either both be antimatter galaxies or both be matter galaxies, and in either case you would never get annihilation.

***Conclusion 4: The advent of immiscibility will also be the time when boundaries between dominia are defined.***

According to the model used, creation occurred symmetrically, although distribution was randomly heterogeneous. Pockets or quadrants could end up with more of one type than the other. These areas will for the remainder of this construct be referred to as: "dominia." Each "dominium" would be expected to initially contain more of one type—matter or antimatter—than the other.

Part and parcel to *Conclusion 2c: The phenomena immiscibility and phase change between gas and plasma are supersymmetric equivalents* is the idea that a characteristic of the process of immiscibility is the phenomenon of sorting. Once divisions between individual particles begin, divisions on an even more macro-scale will also occur because of opposites-repel interactions between neighboring dominia, especially if the "imperfections" are on a larger macro-scale. Boundaries would hence be formed around dominia. This conclusion follows regardless of the original matrix of individual dominia. In fact, it is the position of the Dominium model that boundaries formed around embryonic dominia when these were all still full of other-type material. The distribution of this remnant of other-type material would be similar to the air bubbles in a vigorously shaken bottle of shampoo. Some dominia would be like the air bubbles in the shampoo, while other dominia would be like shampoo bubbles suspended in air. Regardless, with the advent of boundaries, the dominia would then take a form of their own. The fact that primary particles such as protons, neutrons, electrons, etc., are expected to be spheroid solidifies the case. This leads to the conclusion that an individual dominium can be treated as an individual particle. The syllogism is as follows:

a. Immiscibility entails the formation of boundaries.
b. All "imperfections" are particular regions of space of relatively higher density of one type of matter than the other.
c. All boundaries would be drawn along lines of least resistance, i.e. leaving approximately equal density differences within matrix.
d. Therefore, boundaries would form along lines where the matrix of matter to antimatter is close to 50:50.
e. All boundaries will enclose three-dimensional areas.
f. All primary objects will assume a spheroid shape.
g. Therefore, the newly formed objects, dominia, will be spheroid in shape.
h. All other primary objects can be considered particles.
i. Dominia seem to be a primary object.
j. Therefore, when analyzing dominia, one can consider them to behave like particles.
k. Suppose all dominia formed as a result of imperfections in matter-type.
l. All boundaries are drawn based on matter-type and gravitational interactions.
m. Therefore, *Premise 8: Dominia are gravitational particles.*

***Conclusion 5: Self-assembly is a supersymmetric equivalent.***

If dominia behave like particles, then things become interesting. If supersymmetry truly is a reality, then this matrix of particles should behave similarly to primary particles that are electrically based. At a primary level (atomic, molecular, and nano) extremely heterogeneously arranged mixtures of electrically opposite materials undergo a remarkable process referred to as "self-assembly" where the particles self-organize into alternating patterns.[Cheng/Ross] This process is especially interesting when one considers that these systems automatically "choose" to assume a more organized configuration than one that is chaotic. This is interesting and important because it defies the fundamental assertions/assumptions of entropy. This fact is of historic consequence, because Einstein was at a loss to come up with a reason for the apparent order and even distribution of galaxies that we can easily view from Earth. If the primary mixture of dominia in the Big Bang fireball behaved like known/documented electrically based heterogeneously arranged primary mixtures, then this dilemma is easily explained: early dominia underwent self-assembly.

Before proceeding, it is important to define the condition that the dominia will find themselves in immediately after their boundaries have been drawn. The boundaries are drawn based on ratios of 50:50 existing between two opposite-type areas of concentration. If the "inflationary" period led to "imperfections" randomly and heterogeneously dispersed, then dominia will form randomly and heterogeneously.

According to Premise 2, electrical systems provide a supersymmetric forecast of the unknown next step. Systems that are based on electrical interactions (ionic, hydrophobic/hydrophilic, block copolymers, or biochemical) that are extremely heterogeneously mixed all undergo the process of self-assembly. Self-assembly is a very peculiar phenomenon where classic assumptions of entropy and chaos are defied—instead of the system becoming more disorganized, the system organizes itself. In the case of dominia, this organization period serves eventually to account for the modern even distribution of galaxies in the observable Universe, as shown in the following syllogism:

a. Assumption/Premise 8: All dominia are a form of primary gravitational particle.
b. Premise 2: All phenomena that occur in electrical systems are directly applicable as a predictor of gravitational behavior, if the conditions are tightly analogous.
c. All heterogeneous mixtures of primary electrically based particle systems undergo the process of self-assembly.
d. Therefore the original heterogeneous matrix of dominia underwent self-assembly.
e. Therefore: The process of self-assembly is a supersymmetric equivalent between electrical and gravitational systems (Conclusion 5).

The embryonic matrix of dominia of matter and antimatter would be expected to undergo the process of self-assembly (an extremely common event where random particles organize themselves into alternating crystalline structures). Self-assembly is evident in the highly organized crystalline structures of ionic media. Self-assembly has also been well documented in amphiphilic surfactant in aqueous media as well as in biological substances.[Wignall] Hydrophobic-type material congregates with like types, and conversely with hydrophilic materials. The relationship between hydrophobic and hydrophilic materials is an example of the "like attracts like and opposites repel" scenario being put forward by the Dominium model for the relationship between matter and antimatter. During the process of self-assembly of amphiphilic surfactants in an aqueous media, electrical interactions and distribution are the driving force. Also, it is important to note, the process of self-assembly takes the system from a situation of absolute chaos into a more ordered arrangement with predictable nanostructures—micelles,[Wignall] microtubuols,[Wignall] and monolayers.[Muller-Meskamp] The foundation of the Dominium model rests on the idea that hydrophobic/hydrophilic interactions represent a supersymmetric equivalent of the occurrence of immiscibility between matter and antimatter. Proximity of opposites would supply the repulsion and likes would supply the attraction. A heterogeneous gravitational mix will, therefore, self-assemble first, before other interactions show their effect. Thus, when the newly formed heterogeneous mix of dominia reaches the point of immiscibility, self-assembly would result in a re-sorting and repositioning of all dominia within the superstructure to achieve the gravitational stability end-result. Once gravitational stasis has been reached, the period of self-assembly ends.

If random heterogeneity resulted in pockets/dominia of matter and antimatter, one would expect the resulting matrix of dominia originally to be completely random. Quickly, however, a rearrangement and symmetry would have been inserted by the system itself through the process of self-assembly. The matrix of random heterogeneous dominia would constitute a primary mixture in a completely random configuration. It would be expected that such a primary mixture, even at the galactic level, would react like other primary mixtures found in nature: atomic, nano, and molecular. All three of these primary mixtures undergo the phenomenon of self-assembly.[(Wignall, Muller-Meskamp, Cheng/Ross)] The studied forms of self-assembly are all driven by electrical interactions within the initial heterogeneous mix, and the self-assembly put forward by the Dominium model represents a supersymmetry equivalent for the process which happened to dominia at the point of immiscibility. No matter how completely mixed the primary mixture, it will arrange itself into predictable alternating patterns. Several patterns of self-organization have been produced and documented in the lab. Interestingly one pattern looks like polka-dots, or, viewed another way...Swiss cheese, i.e., micelles.[Wignall] Self-assembly is becoming well understood, and since it is so predictable, it has become an invaluable tool of the nanotechnologist.[Cheng/Ross]

***Conclusion 6: Therefore, the super-matrix of the modern Universe would resemble the crystalline structure of a common stone at the atomic level.***

If self-assembly occurred, analogous to what has been observed in nanostructure composite materials made from "opposites," then one could deduce an overall crystalline structure. Therefore, a mathematical model would be needed that could encompass the supersymmetric attributes of an extremely heterogeneous mix of matter and antimatter, similar to the relationships, models, and equations used to define a simple crystal. The syllogism would be as follows:

a. Premise 7: The Universe's original matrix contained imperfections in the random heterogeneous matrix at the point of immiscibility.
b. Then the size of resulting dominia would be variable.
c. Premise 2: All phenomena that occur in electrical systems are directly applicable as a predictor of gravitational behavior, if the conditions are tightly analogous.
d. Then the predictor of the direction of the end-result of the self-assembly process would mirror a similar process for an electrical system, which is heterogeneously mixed and comprised of members with a variety of different electronegativities.
e. Stones are made from mixtures of many different element-types, all with differing magnitudes of electronegativity.
f. Therefore, the super-matrix of all dominia would resemble the crystalline patterns of a common stone (Conclusion 6).

Thus a good analogue of the final/current matrix of galaxies would be a common stone. Many common stones started out as molten material in the magma. As such, individual particles would have high KE and would mix heterogeneously, for the most part. As they cool, pressures mount for the individual particles to find a place to reside under conditions with increasingly less KE. A "final" position is assumed, and things seem to get locked in place. This final step represents the phase transition from liquid to solid, or solidification—specifically, crystallization. A stone, beginning as a mixture of many elements on the atomic level, self-assembles into a crystalline structure. Before the stone cooled, it was a high-KE, thoroughly heterogeneous mix of various elements and ions; as it cooled, it self-assembled into a more orderly arrangement. These conditions are supersymmetrically analogous to the conditions faced by the "imperfections" of the early Universe and are therefore a strong predictor of consequential behavior.

***Conclusion 7: The current structure of the super-matrix of galaxies and the crystalline matrix of a common stone are supersymmetric equivalents.***

Because of random heterogeneity of the original matrix, the gravitational presence of an individual dominium will vary. The supersymmetric equivalent is an electrical system whose members vary in their abilities of electric attraction/repulsion (electronegativity). Stones were made in an originally high KE and heterogeneous mix of different-strength ions and atoms first created in the magma of the Earth, which then gradually cooled; supersymmetrically, the Universe is a cooled, originally high KE and heterogeneous mix of different-strength dominia first created in the Big Bang fireball at the point of immiscibility. Both systems are tightly analogous as high KE and heterogeneous, and, therefore, both are expected to be organized into crystalline matrices of alternating types.

***Conclusion 8: Micelles of opposite-type material existed in embryonic dominia/galaxies.***

As mentioned previously, at the point of immiscibility dominia need not have been entirely rid of all the underrepresented "opposite-type" material. Rather, pockets (micelles) of the underrepresented factor could have been locked inside the formed dominia at that point. As the underrepresented factor, they would be expected to behave like the underrepresented factor in imbalanced electrically based systems that form "micelles."

As defined by Webster's: a micelle is "a unit of structure built up from polymeric molecules or ions." One way to envision them is as spheroid structures suspended in another medium, e.g., soapy water or milk. Underrepresented material trapped within an embryonic dominium would have formed into micelles. This model assumes that that is the case: micelles of underrepresented matter-type remained inside individual dominia at the point at which boundary lines were drawn, resembling the air pockets in a shaken bottle of shampoo. Because of immiscibility, annihilations would have ceased occurring. Micelles would be spheroid, as are their supersymmetric electrical equivalents. The formation of micelles is explained through the following analysis:

a. Suppose that all annihilations ceased once the point of immiscibility had been reached.
b. Antimatter remained at this point in the embryonic Milky Way.
c. Therefore, antimatter in the embryonic Milky Way did not undergo any annihilation events.
d. Then groups of antiparticles no longer come in physical contact with matter particles.
e. Premise 2: All phenomena that occur in electrical systems are directly applicable as a predictor of gravitational behavior, if the conditions are tightly analogous, AND hydrophobic/hydrophilic interactions are an electrically driven system that seems to fit the same conditions.
f. Micelles form in electrically driven hydrophobic/hydrophilic cases where an underrepresented factor is mixed heterogeneously into a substance by which it is electrically repulsed.
g. Therefore, micelles of opposite-type material existed in embryonic dominia (Conclusion 8).

### *Conclusion 9: Soon after the advent of immiscibility, micelles of underrepresented antimatter were collapsed into micelle black-holes—The Dark Event.*

Among the list of possible fates of micelles trapped within the embryonic Milky Way, the most calamitous was their collapse of into micelle black-holes (MBHs). The collapse of micelles is such a profound event that it shall be referred to as the Dark Event. Remember an earlier established premise, Premise 4: *"All black-holes form as the result of an imbalance of forces: Gravitational effects favor collapse, while other forces act to prevent collapse."* The period of immiscibility has been shown by this model to have begun very early on, when the distance between "all objects" was still quite tight. Therefore, all types of gravitational forces would still at that time be in a ballooned phase (through not at as exponentially small fractions of a moment in time earlier). Like-like gravitational interactions within each micelle combined would lie in the same vector direction as the gravitational opposite-repulsion forces imposed by the dominium itself.

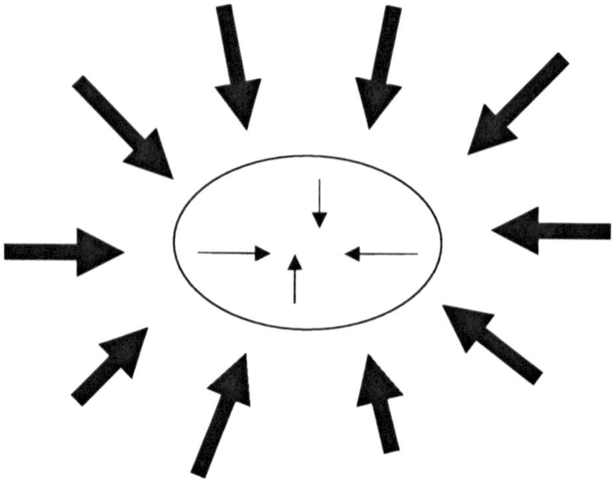

Figure 1: Free-body force diagram of the imbalance of forces on a micelle toward the center of an early dominium, leading to its collapse into an MBH

There would be virtually no gravitational forces acting to cause the micelle to keep its shape. There would still be electrical repulsion forces working to keep atoms apart, but all gravitational vectors would be pointed inward. Gravitational stability has already been shown to trump the system's conflicting tendency for electrical stability. This would cause the micelles to collapse in on themselves and become black-holes of opposite nature to and within the future galaxy. This conclusion is reached, given the assumption that black-hole formation is a function of an imbalance of forces.

An interesting outcome emerges from close inspection of the preceding diagram. The smaller the micelle, the greater the imbalance leading to black-hole formation. Therefore, the smallest of micelles would have the greatest chance of becoming an MBH. As will be discussed later in this work, the collapse of the micelles into MBHs accounts for the sudden decrease in the rate of expansion that is commonly agreed on by the scientific community. Because there is no evidence of a later period of rapid expansion (caused by these small black-holes popping back to their original volume), one can only conclude that small black-holes, once formed, are stable. This assertion is of supreme concern with respect to the very last portion of the scientific model, and will be discussed in Conclusions 41 through 45. Current efforts to create black-hole material in the laboratory rely on the "assumption" that black-holes need to be massive in order to remain stable. At this point in the model, it can be concluded that current assumptions are wrong—dangerously wrong. This cannot be overemphasized, especially when many believe there is no chance to produce stable black-hole material because the masses potentially created at LHC will be "too small."

a. Conclusion 8: Micelles of opposite-type material existed in embryonic dominia/galaxies.
b. Premise 4: All black-holes form as the result of an imbalance of forces: Gravitational effects favor collapse, while other forces act to prevent collapse.
c. The gravitational force vectors felt at any point within each micelle, imparted both by the net matter-repulsion of the surrounding embryonic Milky Way and like-like attraction of the micelle's own antimatter, would lie along the same direction focused inward.
d. **Then the cumulative effect could have been an imbalance in forces large enough to achieve the creation of small black-holes.
e. **Then for all other cases, the cumulative effect could not have been an imbalance in forces large enough to achieve the creation of small black-holes, and therefore that antimatter would have remained diffuse.

Let it be noted that conclusion "d" in this syllogism is not categorical; the path forks. The linkage between premises "b" and "c" is not tight—for some micelles the imbalance could have been large enough to create an MBH, while in others this would not occur. The exact amount sufficient to achieve an imbalance is not currently known. However, it is known that early experiments attempting to create a matter-based black-hole on Earth failed to create a stable black-hole; therefore one can conclude that there is some minimum amount of force imbalance required for the creation of a stable black-hole. Exactly which micelles experienced forces greater than the minimum would have been determined by their position within the embryonic dominium. Micelles toward the center of the new dominium would be expected to feel the largest cumulative forces, while those on the fringes of the new dominium would feel the least. The reason why micelles on the fringes of the new Milky Way dominium would feel the net total force is not captured by premise "c" above. Premise "c" describes the force diagram of a micelle at the exact center of the embryonic Milky Way. For this case only, the vectors will line up perfectly. At the very outer edge of the embryonic Milky Way, the vectors will not line up at all. The micelle's like-like gravitational attraction forces still point inward. However, at this position on the outer edge of the domain, the net repulsion force imparted by the Milky Way would be to expel it to the adjacent antimatter dominium. (See the diagram below.)

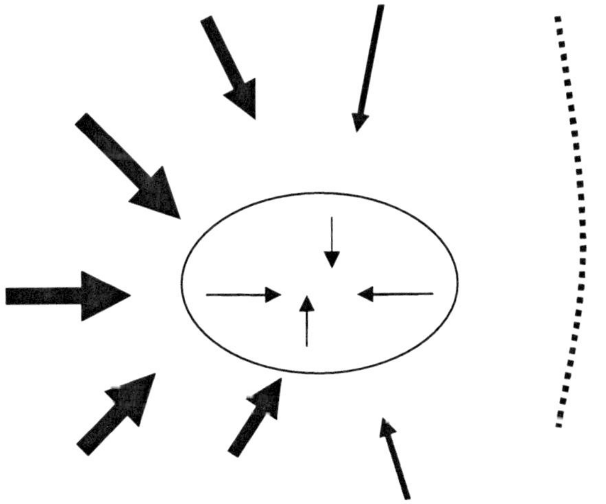

As a result of the skewed gravitational repulsion from the embryonic dominium, the micelles toward the edge would be ejected. Similarly, processes would be occurring across all dominia within the superstructure.

***Conclusion 10: The collapse of micelles into MBHs occurred nearly simultaneously across the entire galaxy.***

Micelle collapse to form MBHs would have happened nearly simultaneously across the entire galaxy matrix. Once boundaries had been established, net force vectors would have also been established. At this point in the analysis there are no known electrically based events that come close to the formation of a black-hole to lend us a supersymmetric prediction of the next events, though it can be shown that the creation of MBHs would have occurred nearly spontaneously throughout all dominia.

a. Conclusion 8: Micelles of opposite-type material existed in embryonic dominia/galaxies.
b. Premise 4: All black-holes form as the result of an imbalance of forces: Gravitational effects favor collapse, while other forces act to prevent collapse.
c. Hypothesis: Gravitational opposites (matter and antimatter) repel; gravitational-likes attract one another.
d. Then net force vectors would be established based on relative geometries of particles in the system.
e. Net force vectors are established as soon as immiscibility boundaries have been established.
f. Therefore, cumulative effects of all net gravitational forces would be felt immediately upon feeling the net-effects/boundaries drawn/immiscibility.
g. All parts of the early Big Bang fireball expanded roughly evenly.
h. Therefore, energy would be diluted in density evenly throughout the expanding post-Big Bang fireball.
i. Therefore, conditions required to meet immiscibility would be met throughout the fireball at roughly the same time.
j. Therefore, the collapse of micelles into MBHs occurred nearly simultaneously across the entire galaxy as immiscibility was established (Conclusion 10).

Because of vagueness of language, a note needs to be inserted regarding the choice of words used to describe the cataclysmic collapse of micelles across all dominia in the Universe and the formation of culvert black-holes (SCGBH) of like material at the centers of galaxies. The word "Dark" was chosen for several reasons. The most straightforward reason is because this event occurred before the formation of stars. At this point in the development of the Universe, there would have been no photons being generated; therefore by definition it would be "dark." (All currently accepted models possess some account for a time before which there were no photons.) The second reason this event was dubbed the "Dark Event" is because this leads to the creation of non-light-generating dense material (MBHs). Later in this model it will be asserted that these objects became migrating structures that help account for what has been observed and recorded as "dark matter." All MBHs, whether composed of matter or antimatter, would never be expected to generate light, and therefore would be "dark." Therefore, the event that led to their creation is aptly named as the "Dark Event." Let it be unequivocally stated that the "Dark Event" is in no manner a reference or homage to anything "evil." According to this theory, the Dark Event was an unavoidable and necessary aspect of the moments after the Big Bang.

***Conclusion 11: The simultaneous collapse of micelles to MBHs had a net effect of pulling apart remaining matter in the embryonic Milky Way.***

When the event of the simultaneous collapse of micelles to MBHs (the Dark Event) occurred across the embryonic superstructure of the Universe, a profound impact on the rate of the overall expansion, dynamics, and energy of the Universe would be expected. To continue the analysis any further, it is necessary to reassess the situation. There are two identified models that could provide some understanding of the next steps. The first model is based on thermodynamics. If across the Universe, micelles suddenly turned into MBHs, one significant effect would be a sudden loss of volume that each micelle once filled compared to the volume of the newly formed MBH. It is conceivable that the Dark Event had a sudden depresssurizing effect on all non-micelle material left in the dominia, thereby pulling apart objects that would normally be gravitationally attractive. If this sudden event were analogous to a thermodynamic adiabatic expansion event, then energy would be removed from the system, consequently assuring permanent immiscibility. The closest example of living through an adiabatic experience would be the theatrical depiction of what happens to items within an airborne plane, when a sudden breach (depressurization event) occurs. Although Hollywood depicts fictional events, or colors up historic ones, for any motion picture there are scientists employed to assure that events in the story mesh with physics as much as possible to maintain illusion of the story. Since movies depicting a sudden breach in cabin pressure show a whipping up of loose items, one can conclude that sudden decompression of the Dark Event resulted in a similar "whipping up" of material. This could lead to a diffuse scattering, or a dust cloud of non-collapsed material. A dust-cloud beginning is the basis of numerous accepted models for the formation of the solar system. Once that dust cloud has formed, the rest of the scenario of how the solar system would be produced is pretty well understood:

a. Recall Conclusion 10: the collapse of micelles into MBHs occurred nearly simultaneously across the entire galaxy. (The Dark Event)
b. When all black-holes collapse they leave behind a void that is filled by adjacent matter.
c. All systems that rush to fill a void with "zero" lapsed time are considered adiabatic.
d. Therefore, the Dark Event was an adiabatic process.
e. Therefore, in terms of the thermodynamic description of the matter left in the Milky Way, the whole event was an adiabatic expansion.
f. All adiabatic events result in a "whipping up" of loose items.
g. Then the remaining matter-portion of the embryonic Milky Way would have been whipped up throughout both the new and old space occupied by the former micelles.
h. Therefore, the resulting matrix would have been less dense, more diffuse, and more like a dust cloud than it had been previously.

A different analysis that leads to the same conclusion is based on the impact on space-time, as described by Einstein. If the classic depiction of a mass's effect to space-time is used—a dimple on the checkerboard continuum—then the rim of that dimple could coincide with the leading edge of the immiscible boundary between that which is micelle and that which is not. As the micelle collapses into a black-hole, that leading edge or rim of the dimple would be dragged inward further toward the micelle's center of gravity. If this happened simultaneously across multiple micelles located toward the center of the embryonic dominium, the checkerboard continuum would have been stretched thin as a result. If space-time was dragged along and stretched as the outer edge of the micelle retracted while black-hole compression was underway, the distances between remaining matter, which would have preexisted in the region of the checkerboard continuum, would have increased.

One example of this process could be the effect seen on the letters of a plastic candy wrapper before and after it is stretched: they get further apart. The syllogism for this model is as follows:

a. Recall Conclusion 10: The collapse of micelles into MBHs occurred nearly simultaneously across the entire galaxy. (The Dark Event)
b. Space-time fills all areas between micelles.
c. All micelle collapses affect space-time similarly by dragging the leading rim edge as the collapse occurs.
d. Then, matter-filled areas not undergoing collapse would feel the effect of space-time being pulled throughout the matrix.
e. In all cases when an elastic object gets pulled, the elongation process causes the object to become regionally less dense as the original mass is pulled over a larger area.
f. Therefore, as space-time was pulled in the remaining matter portion of the Milky Way this remnant matter was pulled apart, becoming less dense.

Although there is no identified supersymmetric indicator for the consequences of the Dark Event, there has been help from understandings of thermodynamics and Einstein's interpretations of space-time. Both models lead to the same conclusion. As mentioned in "Appendix 5: Process and Methodology," two different Aristotelian arguments will always reach the same conclusion as long as the premises are true and the inferences valid. Having two extremely different approaches reach the same answer is itself a source of confirmation of the answer. Therefore, subsequent to the Dark Event, the density of remaining matter in the embryonic Milky Way would have suddenly become reduced as it was stretched/drawn (or whipped up) into the void left by the vanishing micelles. This sets the stage for a diffuse dust-cloud—a prerequisite of most accepted theories of how our solar system was formed.

***Conclusion 12: Micelle collapse would have caused an inflection point in the curve representing the expansion of the Universe.***

If the fate of micelles was to be compressed into tiny black-holes, the point at which this occurred would have been a profound moment—not only for the micelles existing toward the center of each dominium that would collapse simultaneously, nor for the remaining material being whipped up and stretched apart into this newly voided space. The Dark Event would have been a profound moment for the entire Universe. With all this newly available "internal" space, the individual dominium need not expand "outwardly" as much in order to meet demands for expansion set by other aspects of nature. Therefore, the rate of growth for that individual dominium would inflect at this point.

a. Conclusion 10: the collapse of micelles into MBHs occurred nearly simultaneously across the entire galaxy. (The Dark Event)
b. When back holes collapse they leave behind a void that can be filled by adjacent matter.
c. The events described in "a" and "b" occurred while the Universe was rapidly expanding.
d. Then: Void space would have been available for the Universe to continue to expand into internally.
e. Suppose the Universe expanded into the newly voided internal space.
f. Then there would have been less pressure on dominia to grow outwardly.
g. The rate of growth of every individual dominium slowed down.
h. Therefore, the overall expansion rate of the Universe would have slowed down at this point.
i. So: the rate of growth of the Universe slowed because of the Dark Event.
j. Therefore, graphs depicting this sequence of the expansion of the Universe should show an inflection point to mark the influence of the Dark Event (Conclusion 12).

The preceding syllogism provides an explanation for why the rate of expansion precipitously decreased in magnitude at a particular inflection point, as shown on graphs currently used to show Hubble expansion during the earliest moments of the Universe. This inflection point is so severe as to make the graph appear as the assemblage of two distinct periods of time separated by a profound event. Examples of this type of graph are found throughout biology, especially as evidenced by the effects of trends of a cataclysmic event in the fossil record.

One mechanism for the sudden possible decrease in expansion involves the fact that when a black-hole forms, volume decreases precipitously.[Aspinwall,Casadio] The volume-loss created by the formation of MBHs would have been filled by the remaining non-black-hole parts of the fireball. Hence, the rate of expansion of the whole outer volume of the fireball would slow, because internal volume was suddenly available for material to expand into. Therefore, the cataclysmic event that initiated everything was the Dark Event: the advent of immiscibility and the nearly simultaneous collapse of micelles into MBHs.

Another way to describe the inflection point accepted for graphs depicting early Hubble expansion is thermodynamically. The Dark Event has already been described in terms of its similarity to the thermodynamic adiabatic process. In review, the basic law of thermodynamics is:

Internal energy = heat added **to** the system (-) work done **by/from** the system

This author has never liked that formula because of its schizophrenic nature. It attempts to take the perspective of the lab tech at one moment and that of the gas being testing in the next moment. When teaching thermodynamics, a modified equation is much better at showing the true relationships involved in any particular thermodynamic process. By simply picking one perspective and staying with it, the equation seems to simplify, although both versions of the equation are equivalent. The perspective easiest for most to grasp is that of the lab tech. Choosing the perspective of the lab tech implies that the factor describing work needs to be altered. Since the lab tech's perspective is reversed from the perspective of the gas, the equation becomes:

Internal energy = heat added **to** the system (+) work done **to** the system

An adiabatic process is one that happens so quickly, there is no chance for "heat" to leave the system. Therefore the value for heat equals zero. Hence the equation now becomes:

Internal energy = work done **to** the system

From the lab tech's perspective, compressing a system takes work. Therefore, allowing that system to expand is considered "negative (–) work." Therefore, in terms of thermodynamics, the Dark Event will mean a loss of internal energy for the remaining particles of matter. A loss of internal energy would imply a sudden loss of energy to fuel outward expansion. With a precipitous loss of internal energy fueling expansion, the rate of expansion would be inflected at this point.

a. Conclusion 10: the collapse of micelles into MBHs occurred nearly simultaneously across the entire galaxy. (The Dark Event)
b. When back holes collapse they leave behind a void that is filled by adjacent matter.
c. All systems that rush to fill a void with "zero" lapsed time are considered adiabatic.
d. Therefore, the Dark Event was an adiabatic process.
e. Then, in terms of the thermodynamic description of the matter left in the Milky Way, the whole event would be an adiabatic expansion.
f. A loss of total internal energy is a consequence of all adiabatic expansions.
g. Then, there would have been a loss of total internal energy of the matter following the Dark Event.
h. The sudden loss of internal energy meant a loss in energy that could fuel further outward expansion.
i. Therefore, the rate of outward expansion would inflect to match the internal energy loss incurred during the Dark Event. (Conclusion 12, again)

The closest thing to an adiabatic expansion that most readers will know is the phenomenon occurring when hair-spray, bug-repellant, or other aerosols are sprayed out of their can. Because the product is released from the canister extremely quickly, it expands in volume almost adiabatically. As a result of the negative (–) work done to it, the "internal energy" decreases, and therefore, such products always come out cold. Try it. No matter how warm the can might be (though don't heat up the can), the spray always comes out cold. The cooling aspect of adiabatic expansion is the underlying method of refrigeration in a common electric freezer: a contained fluid is allowed to adiabatically expand in proximity to the food compartment, and later outside the food area the fluid is compressed by an electric compressor so that the cycle can repeat. The adiabatic expansion of the remaining portions of the Universe as a consequence of the Dark Event would have had a dramatic chilling effect. Abruptly, and precipitously, everything remaining would cool (drop in temperature) as its internal energy decreased. This abrupt cooling would then be expected to have a profound effect on the subsequent dynamics of the young Universe.

***Conclusion 13: The presence the micelles/MBHs within the early dominial matrix would have a net effect of preventing collapse and creating some of the observed halo structures.***

Whether or not micelles collapsed, they existed within the embryonic Milky Way. Even if the internal pressures of the embryonic dominia were quite large, because of immiscibility, the space occupied by a micelle could not be used as the area involved in the collapse of any matter-type black-holes. The presence of the micelles and their repulsion forces acting three-dimensionally would add cumulative effects to the net gravitational force vectors felt in matter portions of the embryonic dominium. Essentially the effect of this would be to dilute the net force vectors felt in the matter regions of the early Milky Way. Assuming early micelles were inside the embryonic dominia, this solves yet another dilemma: Why didn't early galaxies collapse in on themselves and become giant black-holes? If micelles of an opposite type were present inside embryonic dominia after immiscibility was established, then they'd supply internal repulsion forces protecting development when it was most crucial—when distances were shortest and the ballooned effects of gravity were at a maximum. The expected presence of micelles early on would be expected to have a net effect of "puffing up" the embryonic galaxy analogous to the way vaporizing water and $CO_2$ puffs up baked goods. Similarly, once the micelle is gone, the puffed area would remain.

a. Conclusion 8: Micelles of opposite-type material existed in the embryonic dominia/galaxies.
b. Hypothesis: Gravitational opposites (matter and antimatter) repel; gravitational likes attract one another.
c. Then net force vectors would be established based on relative geometries of particles in the system.
d. All micelles were spread evenly throughout each embryonic dominium.
e. So micelles would be in central regions of the developing dominia.
f. All micelles located in central regions of a dominium would supply components of their gravitational repulsion force in an opposite direction to the net force vector for gravitational like-like attraction.
g. All force vectors that are opposite in direction have the effect of canceling each other out.
h. Then, the resulting net force vector would have been diminished in magnitude compared to its strength had the micelles not been present.
i. Premise 4: All black-holes form as the result of an imbalance of forces: Gravitational effects favor collapse, while other forces act to prevent collapse.
j. Then, regions of matter of the embryonic Milky Way would have been less likely to collapse and form a black-hole than had the micelles not been present.
k. All the immediate areas surrounding micelles/MBHs were protected by immiscibility repulsion forces.
l. Then no matter would occupy these areas.

m. All areas of the galaxy continued to evolve.
n. Areas containing micelles/MBHs would remain devoid of matter.
o. Areas of observed galaxies do show areas apparently devoid of matter.
p. Therefore, both the micellular gravitational force dilution effect and their physical presence within the embryonic dominium would have acted to prevent collapse and led to these observed areas devoid of matter.
q. Hence presence of the micelles/MBHs within the early dominial matrix would have had a net effect of preventing collapse and creating some of the halo structures we observe today (Conclusion 13).

Thus, the presence of opposite-type MBHs within a galaxy would have a net effect of puffing out and preventing the collapse of the entire dominium. Opposite-type MBHs would also be expected to eventually exit the embryonic dominium where asymmetries in geometry exist. Whether or not MBHs have exited a particular galaxy or not, they could have net effects similar to what has been observed since 1952 and described as "halo" object/structures.[Halo...]

***Conclusion 14: Micelles/MBHs with an imbalanced net force would be expected to eventually migrate out of their host dominium.***

As shown, in conclusions 9 through 12, the Dark Event occurred precipitously and cataclysmically. That event had a profound influence on the makeup of the embryonic galaxy. Once the Dark Event concluded, other changes would be expected. As mentioned, micelles located on the outer fringe of the early dominium would have the least likelihood of being compressed into MBHs because the net gravitational repulsion force vector imparted by the matter portion of the early Milky Way would be directed away from the Milky Way's center of gravity. By being located on the outer edge of the embryonic Milky Way, such micelles would also feel a relatively strong gravitational like-like attraction from the closest adjacent antimatter galaxy/dominium. The overwhelming net force vector for such micelles would be to accelerate the particle out of the embryonic Milky Way and toward the adjacent antimatter dominium. As such, micelles and MBHs would migrate, if relative geometries (net force) allows.

a. Conclusion 8: Micelles of opposite-type material existed in embryonic dominia/galaxies.
b. Hypothesis: Gravitational opposites (matter and antimatter) repel; gravitational likes attract one another.
c. Then net force vectors would be established based on relative geometries of particles in the system.
d. All micelles were spread evenly throughout each embryonic dominium.
e. Then some micelles would be in fringe regions of the developing dominia.
f. Net force vectors felt by micelles in fringe regions of the developing dominium would point away from the center of the dominium and toward the adjacent like-type dominium.
g. When net force is directed in one planar direction, acceleration always follows.
h. Therefore micelles and MBHs with imbalance net force vectors will migrate out of the host dominium in favor of an adjacent like-type dominium.

While the presence of these micelles would prevent the embryonic galaxy from collapsing in on itself during the earliest stages of development, those with favorable geometries will eventually migrate out of the dominium. Most of these micelles would eventually be expected to escape the host dominium, much like the bubbles of air in a vigorously shaken bottle of shampoo.

***Conclusion 15: Exiting micelles and MBHs would have "aerated" the infant dominium.***

Conceivably, micelles could remain inside a host opposite-type dominium for a long time if relative geometries allow. In other words, this set of theories does not bar the possibility of micelles of antimatter within the current Milky Way. Conversely, at this point in the discussion these theories do not suggest this possibility either.

Whether or not all the micelles have exited the Milky Way by today, most would be expected to have exited. In exiting, three possible analogies might apply. Either the micelles could have joined to form streams of like material, which would travel antiparallel to opposite streams traveling from neighboring dominia. Repulsion forces would have kept such streams immiscible and no annihilation events would have occurred. Or, traveling micelles could have acted similarly to the air bubbles in shampoo, pushing out unlike material as they travel through followed by the material closing-in in the wake of the micelles. Or, traveling micelles could act in similarly to an earthworm, pushing particles forward and to the sides as they move through. As the micelle passes, although some material might close in behind it, the net effect would be to "aerate" the matrix. None of the three identified possible models of micelle movement would have resulted in an annihilation event once the energy of the system had dropped below the critical level and immiscibility had been established. Of the three scenarios, the weakest is the action of the air bubble in shampoo, because there is no reason to suggest a violation of Newton's First Law: if a particle were accelerated by a passing micelle there does not seem to be any credible argument that it would return to its original position. Therefore, one could surmise that a discrete object, like an MBH, would act like an earthworm pushing material forward and to the sides, while diffuse aggregates of material would travel behind in a stream-like manner. Syllogisms to show stream-like and earthworm-like motion and the effects on the surrounding matter portion of the infant Milky Way are as follows:

**Stream-like migrations:**

a. Net force vectors felt by micelles in fringe regions of the developing dominium would point away from the center of that dominium and toward the adjacent like-type dominium.
b. Premise 4: All black-holes form as the result of an imbalance of forces: Gravitational effects favor collapse, while other forces act to prevent collapse.
c. Therefore, micelles located on the outer fringe of the early dominium would be least likely to collapse and form MBHs.

d. If all such described micelles do not collapse into a black-hole state, then they will remain as diffuse particles.
e. All diffuse particles within a spheroid micelle experience different net gravitational force vectors, depending on their position within the sphere.
f. Then particles with the highest magnitude of net force favoring migration out of the host dominium and toward the neighboring like-type dominium would feel relatively higher rates of acceleration than the rest of the particles.
g. If particles on the leading edge of the sphere (closest to the neighboring antimatter dominium) feel the highest degree of net force, then they will lead the migration.
h. Particles further from the leading edge would be expected to follow behind, but at a slower rate of acceleration.
i. Conclusion 2: Immiscibility: Opposites-repulsion force would lead to a state of immiscibility below a minimum level of energy.
j. Therefore the shape of the micelle would be changed from spheroid into an elongated stream.
k. Assume that at some point the minimum energy needed to sustain immiscibility is reached.
l. So the migration of the contents of the micelles would not lead to any annihilation events.
m. Suppose particles of the former micelles are accelerated in paths that do not allow for annihilation events.
n. Then there must exist a path of least resistance through which the leading edge of particles can pass.
o. All small pieces of mass that might have originally lain along the course chosen by the leading edge particles would feel, at a very far distance away, the gravitational repulsion of the micelle particles as they approach.
p. Therefore, immiscibility forces affect both objects, causing a combination effect where the matter lying in the path of the stream would be repulsed and move while the stream would also feel a repulsion force and would divert slightly.
q. If a gravitational dam is encountered, then all contents of the moving stream would pool in that position.
r. If there is a pooling of antimatter causing an accumulation of antiparticles, the net force imparted on the gravitational dam will rise.
s. So, all imperfections in the gravitational dam will feel increasing pressure from the pooling antimatter.
t. Suppose that one small trickle passes through the gravitational dam.
u. Then the course will be followed and expanded upon as the antiparticles rush past.
v. Therefore, all streaming antiparticles would be expected to eventually leave the Milky Way.
w. Suppose on the other hand that, there exists no route for the "hole" in the original gravitational obstruction to return to its original configuration before being breached by the immiscible antiparticle stream.
x. Then the "hole" will remain vacant after the stream has passed.
y. Therefore, the net effect on the maturing galaxy of migrating micellular streams was to aerate (produce "holes" in) the matter-fabric of the early Milky Way (Conclusion 15a).

As with other forms of streams found in nature, once the path of least resistance is found, it is easy for the rest to follow. Conceivably the infant galaxy could have had areas densely packed with matter that lay across the paths of streaming micellular antimatter. Such densely packed areas could have formed gravitational dams trapping/halting micellular debris from exiting. Such dams would not be expected to hold for very long. Once a crack in the dam formed and was filled by a leading edge of antimatter, more particulate antimatter would follow. The presence of antimatter in this gravitational eddy crack would itself affect the system, much like the root-hairs of a tree can affect fissures in bedrock. As the root-hair swells, pressure builds along the fault line. Eventually the fissure extends through the barrier, and a trickle is allowed to escape. Once the trickle begins, it cannot be stopped. More and more antimatter will rush through the opening causing the opening to become wider. The fissure turns into a hole, then becomes a breach, then becomes a flood. Eventually, all will escape.

**Earthworm-like movement (an alternate mathod/conclusion):**

a. Net force vectors felt by micelles in fringe regions of the developing dominium would point away from the center of that dominium and toward the adjacent like-type dominium.
b. Premise 4: All black-holes form as the result of an imbalance of forces: Gravitational effects favor collapse, while other forces act to prevent collapse.
c. Therefore, micelles located toward the center of the early dominium would be least likely to collapse and form MBHs.
d. All MBHs are compacted to a single point in space.
e. Some MBHs will have net force vectors favoring migration.
f. All MBHs with net force vectors favoring migration would have to move through/past the matter portions of the early dominium.
g. Conclusion 2: Immiscibility: Opposites-repulsion force would lead to a state of immiscibility below a minimum energy.
h. Assume that at some point the minimum energy needed to sustain immiscibility is reached.
i. Then the migration of the MBHs would never lead to any annihilation events.
j. So all migrating MBHs are accelerated in paths that do not allow for annihilation events.
k. Then there must exist a path of least resistance through which MBHs would tend to pass.
l. If a gravitational obstruction were encountered, all migrating micelles would continue to push forward along their paths of least resistance.
m. All matter objects in front of the migrating MBHs would be pushed outward, away from the center of the galaxy.
n. All matter objects located to the sides of a migrating MBH would feel repulsion forces pushing them further to the side.
o. So all matter to the front and to the sides would be pushed increasingly more forward and to the sides.
p. Then the original gravitational obstruction will be weakened and the MBH is able to move forward on its migration.
q. Suppose the process described in n, o, and p goes on indefinitely.
r. Then all obstructions in front of a migrating MBH will eventually be breached creating a "hole" in the original obstruction.
s. Suppose there exists no route for the "hole" in the gravitational obstruction to return to its original configuration before being breached by the MBH.
t. Then the "hole" will remain vacant after the stream has passed.
u. Therefore, the net effect on the maturing galaxy of migrating MBH was to aerate (produce "holes" in) the matter-fabric of the early Milky Way (Conclusion 15b).

Of these three identified modes, the earthworm analogy best describes the behavior of a discrete MBH moving through an opposite-type dominium. This particular construct would be helpful in explaining a few details that have been observed in nature pertaining to our galaxy structure, namely that the outer sections of the Milky Way ring are diffuse while toward the center the structure is more compact. The center of gravitational attraction for the matter early on in the embryonic Milky Way would also be the center of gravitational repulsion for antimatter. As a result, this area would have been one with the lowest traffic of antimatter micelles early on. Micelles would have traveled away from, not through this central point. Conversely, areas furthest from the center of the galaxy would be expected to have been areas with the highest incidence of traffic of both migrating streams of diffuse antimatter and MBHs traveling trough it. Therefore, areas of the outer portions of a galaxy would be necessarily more diffuse, while areas toward the center of a galaxy would necessarily have the highest density of remaining material. These conditions are exactly what is observed in countless galaxies: the central cluster is more dense than outer portions of the galaxy.

***Conclusion 16: Lack of aeration through the center of the dominium led to the creation of a giant black-hole.***

If the migration of micelles did serve to aerate the early galaxy, it is easy to understand the current existence of a giant black-hole at the very center of our galaxy. [NASA#05-166] With just a small influence from micelles to aerate that portion of the developing dominium, that region would remain compacted. Also, being the center of the dominium, this space would feel the highest level of like-like attracttive force focused in one spot. As micelles began to migrate away from the center of the dominium, their effect of preventing the formation of a matter-based black-hole would be diminished. Eventually, forces favoring the creation of a black-hole could rise past the tipping point, and a collapse would occur. The amount of matter involved in this collapse could have been a sizable portion of the original embryonic matrix.

a. Hypothesis: Gravitational-opposites (matter and antimatter) repel; gravitational-likes attract one another.
b. Then MBHs would migrate away from, not through the center of gravity of, each dominium.
c. Conclusion 15b: The net effect on the maturing galaxy of migrating MBHs was to aerate (produce "holes" in) the matter-fabric of the early Milky Way.
d. Then the aeration effect would be felt least at the center of the galaxy.
e. Premise 4: All black-holes form as the result of an imbalance of forces: Gravitational effects favor collapse, while other forces act to prevent collapse.
f. Suppose that the centers of all dominia would be the point of focus for net like-like gravitational attraction.
g. Therefore, the center of the galaxy is the point most likely to generate a black-hole of the same matter-type as the bulk of the forming dominium (Conclusion 16).

***Conclusion 17: Growth of the central culvert black-hole stops as a result of accumulations of micelle-foam at the leading edges of the event horizon.***

Assuming the giant black-hole at the center of our galaxy was formed as proposed above, this process would most likely have occurred during the micelle-rich stage, either immediately after or as part of the Dark Event. Boundaries had just been established and net forces would be realized within the newly defined particle boundaries. Internally directed forces at the immediate center of the embryonic dominium would have been immense, regardless of the presence of micelles. Because of immiscibility, when the collapse occurred, MBHs would be prohibited from annihilating with any of the collapsing material. As the giant central black-hole (in the case of the Milky Way, a matter-based black-hole) was being created, matter would have been drawn in, while any antimatter MBHs, drawn along as space-time stretched, would have been pushed to the perimeter of the event, forming something analogous to the foam that collects in eddy currents on the surface of water in an agricultural culvert. The culvert analogy works well: the water in the ditch would be matter in the galaxy that had not collapsed but was being drawn in; the water will carry with it bubbles (antimatter); the culvert would be this type of black-hole (ours is at the center of our galaxy); and just before the drain occurs, eddies form and the bubbles compile and become a foam. However, in this model, the foam accumulation would act to slow and eventually block the flow of water. If micelle-foam did collect, it would be expected to give an increasingly large repulsive force, slowing more matter that was entering the culvert-black-hole as the micelle-foam accumulated. Remaining non-black-hole matter in the embryonic Milky Way would have felt two main gravitational forces acting on it, being simultaneously been drawn in by the gravitational pull of the growing culvert-black-hole and repulsed by the micelle-foam. At some point, the repulsion force of the micelle-foam would have matched the magnitude of the attraction force of the growing culvert-black-hole and stopped the process. This scenario is consistent with how the galaxy appears today.

a. Conclusion 16: Lack of aeration through the center of the dominium led to the creation of a giant black-hole.
b. The creation of a black-holes causes space-time to be stretched and drags material along with it.
c. So in this process matter would be drawn in close enough to be incorporated as well.
d. Matter that is incorporated into the black-hole no longer takes up volume.
e. Then like-like cumulative gravitation attraction forces will cause whatever was behind the consumed matter to be moved forward.
f. If matter is moved forward, it will eventually be close enough to be incorporated into the growing black-hole.
g. *Conclusion 18: Therefore, this cycle of creating a stable black-hole will never stop if matter is drawn forward continuously.*
h. Also, antimatter (MBHs) would be drawn in closer to the event horizon.
i. Recall Conclusion 2: Immiscibility. Opposites-repulsion force would lead to a state of immiscibility below a minimum level of energy.
j. Assume that at some point the minimum energy needed to sustain immiscibility was reached.
k. Then none of the MBHs pulled close to the event horizon would experience any annihilation events.
l. Suppose MBHs begin to accumulate at the gate of the event horizon.
m. Then it will become increasingly more difficult for matter to pass over the event horizon and into the culvert-black-hole.
n. Suppose the accumulation of micelle-foam continues indefinitely.
o. Then a point will be reached where matter in the infant dominium is no longer able to enter.
p. Therefore growth of the central culvert-black-hole stops as a result of accumulation of micelle-foam at the edge of the event horizon (Conclusion 17).

Step "g" is given status as a separate conclusion in the preceding syllogism because of its enormous importance in understanding the dynamics of black-holes. The importance of Conclusion #18 is that it predicts the dynamic of a black-hole if anyone is ever foolish enough to try to build one here on Earth, as will be shown in the future derivation of conclusions 42 and 43.

If such a stable species were made on Earth, if it bumped into anything, its event horizon would be breached. It would consume. It would continue to bump into things and it would continue to grow as it went along. Gaining mass it would sink into the Earth, consuming any matter it touches. Reaching the core of the Earth, the like-like attractive force would cause matter to be drawn into the black-hole, which would consume at an enormous rate. Days? Minutes? There's no certain way to answer the question, "How much time will it take until the entire planet is consumed into a speck the size of a pea?" One thing is certain: once the process starts, nothing will stop it.

There is no reason to be comforted by the thought, "If it happens, the end will be quick." On the contrary, the process is more likely to be drawn-out rather than quick and painless. In Islamic texts it is said that when the end comes, it will take as long as 80 years for the last of the living to be sucked down to Hell. That time-frame seems to mesh with the scenario of a black-hole situated at the center of the Earth and consuming all that it can. Consider an hourglass: even though all the sand will eventually run out, it cannot rush out because the window for disposal is so small. The same phenomenon would be expected for a black-hole situated at the center of the Earth. Because of three-dimensional competing pressures of individual grains to fall into the hole, the movement of all would be slowed. Therefore, to those at the surface of the planet, the end would probably involve years of absolute Hell on Earth (massive earthquakes, tsunamis, rift volcanoes, and toxic gases) while the Earth slowly and surely caves in on itself. Although many would die during initial traumatic events, many also would be expected to linger and suffer.

### *Conclusion 19: Micelle-foam remains to this day surrounding the giant culvert-black-hole at the center of the Milky Way*

Assuming that micelle-foam stopped the process of the formation of the giant culvert-black-hole (matter) of the early Milky Way, the point at which it stopped would be a stasis point—the gravitational attraction of the central black-hole on the rest of the galaxy would be balanced by gravitational repulsion of the micelle-foam. Because it is a point of stasis, one might easily infer that the system would resist any change taking it out of that stasis, analogous to systems at electrical and other equilibriums found throughout nature. Therefore, it is probable that the micelle-foam would be trapped in its position with respect to our culvert-black-hole and the rest of the galaxy.

a. Conclusion 17: Growth of the central culvert-black-hole stops as a result of accumulation of micelle-foam at the edge of the event horizon.
b. The stoppage of growth of the central black-hole represents a stasis point between all three sets of objects: the culvert-black-hole, the micelle-foam, and the remaining matter of the galaxy which had been drawn in.
c. Recall Premise 2: All phenomena that occur in electrical systems are directly applicable as a predictor of gravitational behavior, if the conditions are tightly analogous.
d. All electrically based systems that reach stasis resist changes to the system that would alter stasis.
e. Therefore, the micelle-foam would be trapped in stasis in its position around the Milky Way's central culvert-black-hole, since it stopped all the way up to today.

The use of central Premise 2 identifies this point in the analysis as yet another example of supersymmetric parallels between gravitational and electrical forces: formation of stasis, or equilibrium.

### *Conclusion 20: Measurements of "Dark Matter" by the Sloan Digital Sky Survey, which appears between current galaxy positions, are in fact indications of migrating MBHs*

Assuming MBHs were created and that some of them possessed asymmetric geometries which caused them to migrate out of our dominium, then, as has been shown, they would have tunneled through the matter of the Milky Way to exit this dominium in order to try to enter an adjoining like-type dominium. While they were in the process of burrowing out of our dominium, the Universe continued to expand. Distances between galaxy clusters increased. Since, during this time, the MBHs were still buried within their host dominium, they too were dragged further and further away from the adjacent dominium to which they were heading. Once the MBHs did escape the matter-based areas of the Milky Way, space-time would continue to stretch between their position and their goal. Although slowed by the stretching of space-time, they would continue to make progress in reaching their goal. It is conceivable that they could be still migrating (between galaxies) toward the adjacent like-dominium, their original goal, along their net vectors of attraction, to this very day. MBHs in the process of making the journey between local dominia would be extremely hard to detect from Earth because they would not emit light and would be between galaxies. Also, their presence would account for the measurement of the "unexplainable" mass existing between neighboring galaxies. This anomalous measurement of material between galaxies has been dubbed "Dark Matter" (Sloan Digital Sky Survey, SDSS) and has been verified by several labs.

a. Recall Conclusion 14: Micelles/MBHs with an imbalanced net force would be expected to eventually migrate out of their host dominium.
b. Suppose the Universe continued to expand while they were still within their host dominium.
c. Then distances between dominia would have increased while the MBHs were still inside their host dominium.
d. Suppose that all expansion of the Universe represents a stretching of space-time.

e. Then the relative rate of approach of the MBHs would be slowed with respect to their goal dominium.
f. Therefore, MBHs may still be in transit today between dominia.
g. Suppose MBHs do exist in transit between dominia.
h. Suppose that all MBHs act similarly to matter-based black-holes.
i. Then they will not be reflective of light.
j. Therefore, they will not be directly observable in the sky.
k. But suppose again that all MBH act similar to matter-based black-holes.
l. Then (2nd), they will gravitationally interact and bend light causing a lensing effect that can be measured.
m. The Sloan Digital Sky Survey has mapped the mass density of the visible sky and located anomalous concentrations of mass located between galaxies and has dubbed this "Dark Matter."
n. Therefore, measurements of "Dark Matter" by the Sloan Digital Sky Survey which appears between current galaxy positions are in fact likely to be measurements of migrating MBHs.

***Conclusion 21: All tests conducted to measure mass-density of the Universe, such as with SDSS, would yield confirmation of the uniformity of mass distribution in all directions.***

Mass-density profiles and background microwave radiation analysis from SDSS and Wilkinson Microwave Anisotropy Probe (WMAP) of the sky, collected by observatories on Earth, appear to show both subtle patterns and uniformity of mass in all directions (excluding the influence of our own Milky Way).[De Bernardis, et al.; Spergel] This observation matches what one would expect from an overall crystal-like universe, assuming we exist deep within this growing/expanding crystal. This conclusion is reached through the following reasoning:

a. Recall Premise 6: During the Big Bang, energy converted into matter and antimatter symmetrically, i.e., in equal amounts.
b. Recall Premise 7: The Universe's original matrix was a heterogeneous mix of "imperfections."
c. Therefore, the original matrix of the Universe contained equal imperfections (matter vs. antimatter) distributed randomly and heterogeneously.
d. Suppose that self-assembly is a supersymmetric equivalent where the heterogeneous mix of dominia moved from a chaotic heterogeneous mix to a more ordered arrangement.
e. Then all effects of self-assembly would occur with the same degree of randomness throughout all regions.
f. Suppose that due to random heterogeneity, all dominia had uneven total masses.
g. Then all dominia would make differing random contributions to the reorganization process of self-assembly.
h. Recall Conclusion 6: The super-matrix of all dominia would resemble the crystalline patterns of a common stone.
i. The distribution of all members was completely random throughout (a rewording of "g").
j. Suppose core sample were to be taken from this "stone."
k. Then all such core samples would have the same degree of randomness and heterogeneity.
l. Then they would all appear to be identical.
m. Suppose all tests, such as SDSS, can be thought of as core samples of the Universe, beginning at Earth and extending into other dominia toward infinity.
n. Therefore, all tests conducted in such a manner would yield conclusions of a uniformity of mass distribution in all directions.

Findings from SDSS confirm the predictions of the Dominium model. The Universe would be expected to be "uniform" in all directions in terms of mass. This appearance of uniformity, however, is a result of randomness and heterogeneity being evenly distributed, not the result of genuine uniformity. According to Dominium conclusions, if you were able to travel in a "straight line" through the Universe, you would cross from matter dominium, to antimatter, to matter, and so on. All of these galaxies within each dominium would vary in terms of total mass present, some being large and some small. Going in a different direction, you would discover a different pattern at first, but as you continue on toward infinity the two samples would converge in their similarity.

***Conclusion 22: The Dominium model is compatible with Einstein's idea of warped space-time, where antimatter has the opposite effect (dimple-up) of matter (dimple-down).***

To begin to approach this question a new premise needs to be introduced: *Premise 9: Parity exists in all forms.* This is a basic premise of the Standard Model. In common English, it means: where there's an up, there's a down; right/left; East/West, positive/negative, attract/repel, etc. In popular graphical representations of Einstein's notions of the warping of space-time, acceleration of particles, and gravitational interactions, space-time is displayed at a giant checkerboard extending in all directions. Objects with mass affect that checkerboard by causing it to dimple. The dimpling effect ripples throughout the checkerboard, though the most concentrated effect is felt at the position occupied by the object in question. The resulting shape can be described as a funnel. In the case of a black-hole, the funnel appears to be very deep (possibly bottomless) and localized. Small objects, such as electrons, would also dimple space-time but their effect would be tied to their size, as is true of all things with mass.

Referring to the classic picture of a planet warping space-time—a checkerboard horizon being dimpled down by the planets effect on space-time[space-time picture]—if you replace the planet with a galaxy you'll produce the same diagram, right? Right, although the net affect to space-time would be a greater dimple. The Dominium model aligns with this assessment. Now how would a neighboring antimatter galaxy warp space-time? Under the current assumption, the antimatter galaxy would also dimple-down the checkerboard. However, that assumption leads to paradoxical conclusions: if neighboring galaxies both dimple-down the checkerboard of space-time, then they should be attracted to one another—yet, that is not what's observed. Reconciliation is found through the Dominium idea of parity: if there is a dimple-down, then there must be a dimple-up. By having the neighboring (antimatter) galaxy similarly affect space-time—the warp-funnel points up rather than down—it becomes easy to see how this relationship results in a repulsive state. Just as two down-facing warp-funnels helps explain gravitational attraction, an up and a down funnel in proximity can show repulsion. Using the analogy of a spring, the restoring force can be in either direction depending on how it's being acted upon. What if lines forming an arch translate into an attract-tion and lines inflecting upon themselves equal repulsion? This solution explains both why chalk falls to the classroom floor and why the galaxies are all getting further apart from each other.

a. Suppose that Einstein's notions regarding matter's dimpling-down of space-time are correct.
b. Suppose that parity exists at all levels.
c. Then within Einstein's model, antimatter will cause space-time to dimple-up.
d. Suppose that Einstein's notions of how two dimpled-down regions of space-time explain gravitational attraction between two particles of matter are correct.
e. Assume that parity exists at all levels.
f. Then, all examples of two dimpled-up regions of space-time will be a correct explanation of gravitational attraction.
g. Assume again that parity exists at all levels.
h. Then, also, a system with one dimpled-up component and one dimpled-down component will gravitationally repel.
i. Assume that the Dominium model's central hypothesis (opposites repel and like attracts like) always aligns with "f" and "h."
j. Therefore, Conclusion 22: The Dominium model is compatible with Einstein's idea of warped space-time, where antimatter has the opposite effect (dimple-up) of matter (dimple-down).

### *Conclusion 23: The event horizon will be flat.*

Assuming that matter and antimatter warp space-time in an opposite way, then in terms of space-time's warping of galaxies, one would expect a flat overall event horizon.[De Bernardis et al.] This conclusion comes from the following: if the "funneling" flows opposite for antimatter galaxies as represented on a classic planar illustra-tion, then the crystalline regular alteration of weight would stretch out to be flat on an even larger macro-scale. This phenomenon would be similar to that of the assembly/physics of the secondary structure of sheet proteins, resulting from equal amounts of charge interaction being situated on one side of the "sheet" as the other, thereby stretching it flat. If the preceding assessment were correct, then it would represent another manifestation of supersymmetric relationship between gravitational and electric forces. Test results seem to confirm this and indicate a flat horizon as recorded through analysis of cosmic microwave background radiation[de Bernardis] and has been supported by further analysis of SDSS and (WMAP) results.

a. Recall Premise 6: During the Big Bang, energy converted into matter and antimatter symmetrically, i.e., in equal amounts.
b. Recall Premise 7: The Universe's original matrix contained imperfections and was a random heterogeneous matrix.
c. Therefore, within any plane there will exist an equal amount of antimatter versus matter.
d. Recall Conclusion 22: The Dominium model is compatible with Einstein's idea of warped space-time, where antimatter has the opposite effect (dimple-up) of matter (dimple-down).
e. Then the net warping effect on space-time will be zero.
f. So suppose there is no net warping effect on space-time within any plane.
g. Then the event horizon will be flat.

## Part 3: The Dominium Lens for Current and Future Manifestations

This section will explore ramifications of the Dominium model, as they exist in our current and future Universe. All premises and conclusions from the first part of the analysis will be assumed to hold their validity throughout this portion of the analysis. Prior premises and conclusions will be used as tools to construct arguments and make predictions.

The common model of our solar system is well understood: our Sun at the center; roughly nine planets, of which Earth is number 3; an asteroid belt; and a distant belt of icy objects (some might be large) through which comets journey as part of their elliptical orbits. Our position gives us "Goldilocks" conditions for life: the radiant energy from the Sun yields surface temperatures along the scale of liquid water.

What is not understood is vast. A few of these remaining questions can be answered using the Dominium model. The driving force behind the solar wind, data collected from solar wind monitoring satellites, and the magnetic twisting observed on the solar surface are all predicted by, and their existence supports the validity of, the Dominium model.

This section will outline and define the role played by positrons in the generation of the solar wind, given the assumptions of the Dominium model that matter and antimatter induce a repulsive force on each other. One function of this section is to review recent research on equipment and tests designed to show that currently it is impossible to measure the titer of positrons within the macro solar wind plasma structure. Also, this part of the work will expand on the consequences of the solar wind's cumulative effects on the Universe, and on how these effects set the stage for an eventual imbalance of forces within the Universe's macrostructure leading to predictable collapse—the next Big Bang.

Current theory regarding the solar wind states that the "solar wind forms as a result of the competition between the cold vacuum of space and the hot dense corona."[NASA public information website] However, which specific mechanisms cause this process cannot be agreed upon. The solar wind is known to be a plasma of high-energy particles (uncoupled) moving at mach speeds.[NASA ACE] It is generally accepted to be a unidirectional flow of matter—a paradoxical conclusion that seems to defy understood gravitational attraction and principles of kinetics, yet one that matches all recorded data of this phenomenon to date.

NASA has several programs currently working to uncover the mystery of the solar wind. Why does it move in a unidirectional flow from the Sun? What underlying force is at play? Where do the particles stop, or do they even stop? And what is the solar wind composed of?

The programs have machinery currently taking data. NASA has listed on its websites the Experimental Overview, Science Objectives, Measurement of

Objective, and Descriptions of Instruments for all the different detecting devices for both missions. However, throughout the multitude of things being tested, the presence and concentration of positrons is not listed. Nor have samples been detected.

Of NASA's listed programs, SWEPAM, SWE, and 3D-PLASMA, all study particles within the expected range of positrons (1ev to 400keV)[NASA ACE]. All three list the use of detectors as part of their product design. SWEPAM is the most specific, listing CEMs (channel electron multipliers) to detect electrons.[McComas] SWE lists the use of "silicon detectors," while 3D-PLASMA describes "detectors." [NASA ACE] It's safe to assume that all projects are using CEMs, since CEM technology was developed, perfected, and proven reliable under the harshest conditions in labs such as CERN.

CEMs are employed in a variety of applications and their technology is well understood. CEMs employ solid-state technology in a wafer design. N-layers and p-layers are laid out with the keenest precision. When installed they are designed to cascade and send an electric pulse when influenced by a passing particle of a particular range. The drawback of CEMs is that they cannot differentiate between particles. Manufacturers of CEMs advertise the same product to be used for the detection of positrons as well as electrons. CEMs' efficiency to pick up the presence of positrons is well documented.[Karar] Therefore, the CEM readings obtained from both WIND and ACE for "electrons" could easily include hits with positrons as well. To avoid confusion, researchers utilize faraday cup analyzers to modulate electric fields to accurately pinpoint different ranges within the particle spectrum for analysis. Although the sampling devices on WIND and ACE could pick up evidence of a positron if one were to pass through the device, no such occurrence has ever been measured.

The lack of recorded evidence of positrons within samples of the solar wind has left many to conclude that positrons play no role in the dynamics or generation of the solar wind. The Dominium model takes a different stance and will show how the positrons are, in fact, the most important players in the dynamics and generation of the solar wind. The lack of a record of positrons passing through any of the detectors of the monitoring probes concurs with predictions of the Dominium model. But before that can be deduced, the fundamental principles of this phenomenon need to be established.

***Conclusion 24: Positrons produced in the core of Sun will be accelerated toward the surface.***

It is well understood that the process of fusion occurs within the Sun. This high-energy process takes small atoms, fuses them together, and releases energy. During this process antimatter, in the form of positrons, is produced. For every one helium-4 atom that is produced, two positrons must also be created. This is a fact that is universally accepted within the scientific community. Based on the energy output of our Sun, literally tons of positrons are created within its core each second—this fact is also universally accepted. What happens next is up for debate. Some theories conclude that this large volume of positrons all annihilate within the core of the Sun and contributed to the Sun's energy output. Some theories hold that the positrons migrate and eventually annihilate in the mantle or corona of the star. The Dominium model takes a different stance than those theories that preceded it: some annihilation events occur in the core; positrons migrate; and some escape.

According to the Dominium model, opposites repel. Positrons produced in the core of the Sun would feel a repulsion force exerted by the Sun itself as soon as they come into existence. As a result, they begin their migration toward the surface of the Sun.

a. All positrons produced in the Sun are antimatter.
b. Hypothesis: Gravitational-opposites (matter and antimatter) repel; gravitational-likes attract one another.
c. Therefore, all positrons would be repulsed by the matter of the Sun.
d. Therefore, all net force vectors for all positrons would point away from the center of the Sun.
e. Suppose that F=ma, Newton's $2^{nd}$ Law, is always true.
f. Therefore, the positrons will be accelerated toward the surface of the Sun once they are produced.

***Conclusion 25: The need for a system to gravitationally sort is the primary driver in cases of absolute chaos, such as the Big Bang.***

This conclusion was in fact already implicitly shown in previous discussions and syllogisms used in this model, though it was not explicitly delineated. The conclusion that immiscibility occurred very early on within the initial stages of the Big Bang is evidence of the fundamental priority of the gravitational force to all other forces. The common understanding of gravity as it manifests on Earth is that it is a "relatively weak" force when compared to electricity. However, such assessments are made within a system (Earth) that is uniform in makeup (all matter); therefore in terms of gravity it is not "out of balance." Rather, the like-attracts-like interactions that are observed are occurring in a stable gravitational system. Supersymmetrically speaking, because like-particles have been sorted together already, modern Earth conditions would be equivalent to the "coupled" state for electrical charges. Because the lower energy state has already been reached, there is no huge imperative to change anything drastically. Still considering supersymmetric parallels, coupled atoms do interact with other coupled atoms, as is the case in chemical bonding. However, the energies involved with chemical bonding are much lower than those involved in the coupling/decoupling process. Similarly, the energies displayed in gravitational interactions on Earth will be much lower than those involved with immiscibility/miscibility. Also in analyses above, it was shown that micelles existed with embryonic dominia at the time of immiscibility.

What was not mentioned in previous analyses of those events is that there was an alternative option—no micelles existing within embryonic dominia. This path was ignored because only the presence of micelles in the embryonic dominia could have lead to Conclusion 13: The presence the micelles or MBHs within the early dominial matrix would have a net effect of preventing collapse and creating some of the observed halo structures. Since we know that the early Milky Way didn't collapse in on itself and since we've observed halo structures apparently devoid of mass, the route containing micelles early on was the only valid path. Given that micelles had to be present very early on to result in the Milky Way to account for its current configuration, the hierarchical drive to achieve gravitational sorting (immiscibility) is primary, and prior to the other known forces.

a. Recall Conclusion 2c: The phenomena of immiscibility and phase change (coupling) between gas and plasma are supersymmetric equivalents.
b. As energy leaves a plasma system all aspects of the process of changing states from a plasma to a gas involve electrical sorting.
c. Therefore, as energy leaves a miscible system all aspects involved in attaining immiscibility involve gravitational sorting.
d. Immiscibility is always an immediate event to occur in a Big Bang fireball.
e. Therefore, gravitational sorting is always the first event to occur in any completely chaotic heterogeneous system.
f. The order in which physical events occur is an indicator of hierarchy.
g. Therefore, the need for a system to gravitationally sort is the primary driver in cases of absolute chaos, such as the Big Bang.

Because the positrons, predicted to be accelerating toward the surface of the Sun, will be entering areas of increasingly lower densities of matter as they climb higher and higher into the corona, eventually particle density conditions will be met for immiscibility to reign. At the point at which conditions for immiscibility have been met, the mixture of positrons versus forms of matter will be extremely heterogeneous. Since conditions for immiscibility have been met, and because the matrix is extremely heterogeneous, the system will be expected to immediately undergo the self-sorting process inherent in immiscibility. The positrons will be compartmentalized together despite the fact that they feel electrostatic repulsion between each other. The Dominium model shows that there are times when gravitational/space-time forces will be dominant in the dynamics of the system—cases of extreme heterogeneity of matter and antimatter in the same local region. In such cases, the system will self-sort based on gravitational-type, before being driven by any other forces because of hierarchical positions. During this period, electrical forces would be relatively mild players in comparison to the effects imparted by gravitational interactions. Once self-sorting is completed and immiscibility established, fundamental gravitational requirements will have been satisfied. Once immiscibility is complete, electrical interactions will take over as the dominant force within the micelle/dominium.

*Conclusion 26: Some positrons will escape the Sun.*

It is known that a byproduct of fusion is the creation of positrons. In order to create one atom of Helium-4, two positrons must be produced. The number of positrons being produced each second in order to generate the measured continual output is enormous. As the positrons begin to accelerate away from the core of the Sun, they will be traveling through matter-dense areas that become progressively less dense. As they climb higher, they would continue to accelerate and gain kinetic energy. Both the added kinetic energy and the fact that the antiparticles are moving through matter-dense areas favor miscibility and therefore annihilation events. However, the distance to open space (conditions favoring immiscibility) is not infinite. Also, the superheated gases at the upper layers of the Sun would be less densely packed. Therefore there is a chance, albeit slim, that an individual positron could reach a position where immiscibility would establish itself. Given the huge volume of positrons being generated each second, the likelihood that some positrons will escape grows to the point of being probable.

a. Positrons produced in the core of Sun will be accelerated toward the surface.
b. The need for a system to gravitationally sort is the primary driver in cases of absolute chaos, such as the Big Bang.
c. Therefore, as they accelerate toward the surface all positrons will attempt to group with other positrons.
d. Group formations may enhance the ability of a minority of positrons to reach the surface.
e. Suppose a minority of positrons reaches the surface.
f. Then that minority will escape.

This particular portion of the model can be illuminated through metaphor. Imagine "water" beginning at a "source." Beginning at the source, the water must flow over ten miles of extremely absorbent, unquenchable sand aquifer. Ten miles away is the "garden" toward which the water is traveling. If a little stream flowed, once it hit the sand it would survive for five feet or so before being swallowed up. Change the stream into a river, and it survives 500 feet before being consumed. Put the flow of the Nile and the Amazon together and it lasts a mile before being absorbed. Increase the rate and eventually there will be some flow at which the ten-mile stretch of sand will be traversed by some water and reach the garden. In the preceding model:

1. water = positrons
2. source/flow = rate of fusion in the core of the Sun
3. 10 miles sand = distance needed to be traversed to attain immiscibility
4. Absorption = Annihilation loss of positrons attempting to make the journey.
5. Garden = The point at which conditions for immiscibility have been met
6. Suppose the rate of water/positron flow is high enough.
7. Suppose the production of positrons in the solar core truly is enormous.
8. Then, some positrons will escape out into free space (to a position where immiscibility is yet to be established).

***Conclusion 27: Immiscibility, and the formation of micellular positron packets, occurs within the Sun.***

The position where immiscibility could begin need not be outside of the coronal region. In other words, that "garden" might not necessarily be in free space. According to earlier analysis (mainly conclusions 2, 3, 4, and 25), the drive for a system to become gravitationally stable is the first-priority driver of a completely chaotic mixture, as happened extremely early on in the Big Bang fireball sequence of events. Conditions within the Sun are harsh, yet the Big Bang fireball would be expected to be even more severe. Therefore, the primary tendency for the sorting involved with immiscibility could be expected to occur within the Sun, not in free space.

a. Suppose that within the core of the Sun there is always a chaotic heterogeneous mix of material: plasma of matter and exotic particles being produced, esp. antimatter positrons.
b. Recall Conclusion 25: The need for a system to gravitationally sort is the primary driver in cases of absolute chaos, such as the Big Bang.
c. Then, in the Sun, positrons would immediately begin grouping together to form micellular packets.
d. Recall Conclusion 26: Some positrons will escape the Sun.
e. Therefore they will be escaping the Sun as micellular positron packets.
f. Also, therefore, they traveled up through the Sun in the form of micellular positron packets.
g. When conditions for immiscibility had been achieved, all positrons would be clumped as micelles of similarly accelerated particles.

Therefore, at some point away from the core of the Sun, escaping positrons will self-assemble into micellular positron packets (MPPs) of similarly accelerated particles. The supposition that even within the Sun, conditions will be met where immiscibility prevails is not without merit. The Dominium model originally predicted that in the earliest moments of the Big Bang, immiscibility conditions were met (total regional energy and particle density dropped below a certain level). Conditions within the Sun, although harsh, don't necessarily meet the minimum requirements. There is even reason to argue that MPPs might begin to form beneath the corona. However, any line of argument, at this time, is merely speculation, because the exact formula-mix of conditions has not yet been defined and therefore cannot be accurately applied to a structure such as Helios. For the sake of this work, it will be assumed that the conditions necessary for immiscibility do occur somewhere at a distance from the core of the Sun.

***Conclusion 28: The solar wind is the result of the physical interaction of immiscibly protected MPPs with ions and electrons in the upper atmosphere of the Sun.***

It has also been put forward that once conditions for immiscibility have been met, antimatter will move through matter like an earthworm or a stream, pushing it forward and to the sides. The actions of the MPPs could also be a combination of the two forms of transport, since they would begin as more discrete entities and eventually be expected to coalesce into a stream. If positrons are escaping and pushing matter at the uppermost atmosphere of the Sun forward (away from the Sun) and to the side, it is easy to imagine that this continual nudging will accelerate small particles, such as ions and electrons, away from the Sun. Hence the solar wind:

a. Recall Conclusion 15: Exiting micelles and MBHs would have "aerated" the infant dominium, via earthworm-type movement pushing matter forward and to the sides.
b. MPPs are also three-dimensional micelles.
c. Then MPPs would also affect matter in the upper atmosphere by pushing it outward and to the sides.
d. Suppose that $\sum F = ma$, Newton's 2nd Law is always true.
e. Then the matter in the upper atmosphere of the Sun (uncoupled ions and electrons) would always feel a net acceleration away from the core of the Sun if enough nudging were applied by MPPs.
f. Suppose that the steps "d" and "e" repeat continuously through continuous fusion in the Sun's core.
g. Then, the acceleration imparted on uncoupled ions and electrons in the upper atmosphere could be quite large.
h. Suppose a large acceleration would always be needed for matter to overcome the gravitational attraction of the Sun.
i. Then, some matter might be accelerated enough to resist the gravitational attraction of the Sun.
j. All matter accelerated in this manner would be expected to be in a unidirectional flow away from the Sun.
k. The phenomenon known as the solar wind is a unidirectional flow of uncoupled ions and electrons away from the Sun.
l. Therefore: The solar wind is the result of physical interactions of immiscibly protected MPPs with ions and electrons in the upper atmosphere of the Sun (Conclusion 28).

In the core of the Sun, the process begins as a mixture of positrons, electrons, and ions. The positrons immediately begin to "school" and move away from the core of the Sun. Like-like gravitational attraction between positrons would coalesce into structures of micellular schooling positrons en route to the solar surface. Initially, annihilations would occur between matter and positrons on the outer fringes of a given school of accelerating positrons. Further coalescing could promote the survival chances; in the same way, schooling increases the survivability of an individual within a given school. At some finite point, conditions for immiscibility are reached and MPPs, the schooling positrons, would no longer annihilate with any of the surrounding heliosphere plasma. These MPPs would push upward on the matter of the heliosphere above them, because they would now have an immiscible three-dimensional structure and are still being accelerated through gravitational repulsion against the center of mass of the Sun. Matter highest in the atmosphere would be expected most likely to be accelerated past its escape velocity. Matter with velocities high enough to escape the pull of the Sun will continue to be pushed forward and to the sides by successive MPPs emerging from the solar surface, though at a continually lower intensity as distance from the source increases.

The expectation that the MPPs would merge, despite being composed of all positive particles, is based on the premise that immiscibility was in effect during the extremely harsh conditions immediately after the Big Bang. The system's inherent requirement for gravitational stability (sorting via self-assembly) trumps the system's secondary need for electrical stability. Therefore, despite the fact that the creation of MPPs sets up an electrical imbalance, it must occur in order to satisfy conditions of immiscibility and self-assembly to achieve gravitational stability. Once the MPPs are formed, the particles within would be expected to behave like a school of fish, though loosely connected because of like-like electric repulsion. Like fish within a school, the positrons would be affected by stimuli similarly, causing them to travel together, altering course in unison in response to ripples in space-time, changes in charge, and shifts in magnetism in their paths. The MPPs would be expected to divide to avoid obstacles, only to reform on the other side.

***Conclusion 29: The magnetic fields on the solar surface will be complex, strong, and intertwined.***

One fact is certain: the movement of positrons away from the core of the Sun will generate individual magnetic field lines, based on the elementary formula $F=qv\beta$. This would be true of all positrons migrating radially out in every spherical direction. If, prior to conditions of immiscibility, positrons did begin to school and form MPPs, then field lines generated by a school would become locally stronger because of cumulative effects. A problem with this situation will arise quickly: because positrons/MPPs are traveling to the surface in all directions at once and generating magnetic fields as they go, the generated fields would compete with each other. This may slow them all down due to competing Lorenz forces, despite the large gravitational repulsion from the Sun. Being slowed down would decrease the kinetic energy and therefore would favor conditions of immiscibility. Regardless, the positrons/MPPs would be expected to push forward generating competing magnetic fields as they go. This competition between generated field lines may translate into the twisting of magnetic fields at the Sun's surface as documented/evidenced in numerous observations.[Steele Hil]

a. $F = qv\beta \sin\theta$ (force equals the product of the perpendicular components of charge, velocity, and magnetic field) is always true.
b. All positrons generated in the core of the Sun have a charge of $e^+$ and are moving at a velocity due to gravitational repulsion from the Sun.
c. Therefore, all such positrons will be generating magnetic fields as they travel.
d. Suppose positrons are moving away from the core radially in all spherical directions.
e. Then the pattern of field lines will be complex in all spherical directions.
f. All MPPs tend to coalesce as they travel.
g. Then, the net charge of a formed MPP will be the sum of the two previous micelles.
h. Therefore, the strengths of all magnetic field lines generated by individual MPPs traveling through the Sun will be strongest toward the end of the journey (at the solar surface).

i. So all MPPs emerging from all portions of the solar surface will be generating their strongest magnetic fields.
j. Therefore, the magnetic fields on the solar surface will be complex, strong, and intertwined (Conclusion 29).

Multiple observations of this predicted phenomenon have been recorded. Solar flares twist and spiral, actions attributed to complex twisted magnetic field lines. Swirls and even "coronal mass ejections" (CMEs) have their roots in the strong, complex, and intertwined nature of magnetic field lines on the solar surface. As they grew in size the MPPs would also grow in their affect/force upward on the matter of the outer reaches of the Sun. Perhaps large coalesced micelles of anti-matter positrons attempting to escape the gravitational influence of the Sun could play a role in lifting up the largest sections of mass off the solar surface. Once free of the surface of the Sun, the MPPs would continue to merge, and to push matter forward and to the sides. The matter being pushed would be accelerated a little more with each push.

***Conclusion 30: Space probes designed to measure the solar wind will never detect the presence of positrons.***

***Conclusion 30b: Any matter located in the path of the solar wind would act as a conduit-seed for matter-streams of solar wind to flow past.***

Problem: Why haven't these streams of positrons been measured by WIND or ACE or any other mission sent up by ESA, or JSA, or anyone else? The reason is the classic paradox of the anthropologist in the hut trying to study the "normal" behavior of the natives: his very presence in the hut changes the behavior of the natives, yet how can he study the natives if he is not there? This is exactly what is happening as a result of the test probe being situated in that spot in space. The space probe is the anthropologist. Being made of matter, our instruments would also be expected to be permanently immiscible with the antimatter streams, since conditions for immiscibility were met much closer to the solar surface. The space-craft, a matter-dense object, would affect space-time. If the effect on space-time caused by the presence of matter is to send out "dimple-down" ripples for a very far distance, the MPPs would feel the effects of the space-craft on space-time at a very large distance away, while matter streams accelerated by the MPPs would also feel the spacecraft's gravitational presence. Matter steams would be drawn to the probe, while antimatter streams would be repulsed This assessment requires only one premise to hold true: gravitational effects must be felt at extremely large distances, far enough away as to give the high-speed antimatter time to alter its course. For the anthropologist, the true goal is to figure out what is the normal dynamic based on the dynamic that he records, which has been changed by his presence. For the scientist, that would mean an acceptance that the matter-based portion of the solar wind is what's presenting itself. From the hints and clues buried in this data, what can be surmised of the MPP portion of the solar wind?

*a.* Recall Conclusion 28: The solar wind is the result of physical interaction of immiscibly protected MPPs with ions and electrons in the upper atmosphere of the Sun.
*b.* The presence of any matter will cause space-time to dimple down.
*c.* All changes in the ripples of space-time are felt at very large distances.
*d.* Therefore, both the positrons and the matter (ions and electrons) in the solar wind would feel the presence of the matter-dense space probe from a very large distance away.
*e.* Hypothesis: Gravitational-opposites (matter and antimatter) repel one another; gravitational-likes attract one another.
*f.* Therefore, matter will alter its course due to like-like gravitational attraction and positrons will alter their course due to opposites' gravitational repulsion.
*g.* All space probes are matter-dense objects.
*h.* Therefore, all positrons will react to avoid coming within proximity of space probes.
*i.* Therefore: space probes designed to measure the solar wind will never detect the presence of a positron (Conclusion 30).
*j.* Any matter located in the path of the solar wind would act as a conduit-seed for matter-streams of solar wind to flow past (Conclusion 30b).

Thus, the lack of any recorded evidence of positrons within the data collected by WIND or ACE is predicted by the Dominium model. Its absence lends support to the idea that positron escape is the cause of the solar winds and to the fundamental premises that opposites repel and that the effects on space-time by any mass are felt at extremely large distances.

While the spacecraft would repulse the MPPs, the matter would be attracted. The presence of the spacecraft would act analogously to an imbalance of dopant material in a defective CEM or other solid-state capacitor. The CEM is now "defective" because cascades now occur primarily through the imbalanced portion of the wafer.[Karar] Therefore, the WIND craft, itself, would act as an anchor for the matter streams (even when the titer of matter happened to be temporarily low in local space at that moment), while the antimatter would be expected to flow in adjacent space. Adjacent might not necessarily mean close by. The cross-sectional area of the matter stream using the spacecraft as a conduit could be quite sizable. Any attempt to pierce the boundary, or shoot something through the boundary, would be accompanied by a subsequent shifting and/or movement of the boundary to meet the conditions of the altered system. Any matter-based projectile would instantly become another seed-conduit for another matter-based plasma stream. Also, because of immiscibility, if the wave-front of dense MPPs passes by, any matter in the matrix of the solar wind would still be the only particles passing through the near-space surrounding the spacecraft, as this is its path of least resistance. Even if virtually no matter was in the mix of a wave front, MPPs would still have felt the gravitational effect of the space station from far off and would still avoid passing near it, leaving the space immediately around the craft seemingly free of traveling particles, causing counts recorded by detectors to decrease dramatically. Hence, a period of lull in particles measured by a space probe would correspond to a period when MPPs are passing by in large number. A spike would follow this interlude in measurement. After the MPP front passes, it will be followed by another front of mass, and when this hits another spike occurs.

This would account for the sharp boundaries observed in solar wind plasma structures.[Riazantseva]

***Conclusion 31: Measurements of matter in the solar wind should show a predictable pattern of ebbs and flows consistent with concentric fronts radiating outward and passing the probe.***

There is constant variability in the solar wind. When looking at data retrieved from WIND and ACE, we find that the solar wind, although continuous, goes through predictable ebbs and flows. This observation can also be explained by the Dominium model. If positrons are accelerating away from the core, and if immiscibility is the normal condition of matter and antimatter under a certain level of total energy and density of material, then some positrons will reach such conditions and become permanently immiscible and free to eventually journey outside of the Sun. The chance for escape would be possible from all sectors of the solar surface; however, some paths for escape would likely be easier than others. As a result, one would expect some areas with relatively high release of MPPs, while other areas would be comparatively low. This conclusion seems to be backed up by work that has measured solar wind flow and correlations to the appearance of coronal holes unevenly distributed on the surface of the Sun.[Temmer] If there is a variation in the ability of MPPs to leave the surface of the Sun at a given moment, then there would be a heterogeneity in the spray pattern of MPPs from the solar surface. Because the Sun is rotating, the spray of MPP would emanate and form a pattern much like a rotating garden sprinkler, with successive fronts radiating outward. For the solar wind, this would mean alternating fronts of matter mostly followed by antimatter. Matter is being pushed forward and away by the micelles of antimatter behind it. Both fronts are radiating out and away from the Sun.

a. The Sun is a heterogeneous composite of materials.
b. Suppose MPPs are forming and accelerating through this heterogeneous body.
c. Then some paths will be less likely than others to incur collisions.
d. Then some paths will lead to areas at the surface with relatively larger quantities of MPPs.
e. Recall Conclusion 28: The solar wind is the result of the physical interaction of immiscibly protected MPPs with ions and electrons in the upper atmosphere of the Sun.
f. Therefore, in all areas with a higher proportion of MPPs will have a greater net effect on matter present than will other areas of the solar surface.
g. A greater net effect always translates into a larger net mass possibly pushing into space.
h. Then all areas with the greatest release of MPPs would be accompanied by the largest amounts of matter being pushed forward.
i. The Sun is always spinning.
j. So the areas of maximum matter being pushed forward will always spin with it.
k. Devices that spray out a stream of particles while rotating form patterns of concentric fronts radiating outward.
l. Therefore, the solar wind will appear as a pattern of concentric fronts radiating outward.
m. But matter-based probes like WIND or ACE will only measure the matter being carried along.

n. Therefore, measurements of matter in the solar wind should show a predictable pattern of ebbs and flows consistent with concentric fronts radiating outward caused by the matter passing the probe.

***Conclusion 32: Matter portions of the solar wind will gravitate toward paths that extend outward via the interplanetary disc and will be drawn to and hop from planet to planet when alignments allow.***

According to the predictions of the Dominium model, any/all matter orbiting the Sun would act as a conduit for streams of matter-dense material. Conversely, antimatter, e.g. micelle-black-holes, repels matter and attracts MPP streams. Planets impart a large effect on space-time because of their mass, and hence would act as major gravitational conduit structures for matter being pushed along by the micellular positrons of the solar wind. Therefore the streams of matter would tend to skip from planet to planet when alignments allow. This phenomenon has been recorded, although the phenomenon is commonly attributed to the presence of interplanetary magnetic fields.[NASA Public Info...] If all objects accelerate along paths of least resistance, and since most matter within our solar system exists within the interplanetary disc, then streams of matter pushed by the MPP would be more likely to travel within this disc, while MPP would be more likely to seek paths that do not go through the disc. Even though the MPP and matter portion of the solar wind would not travel through the same space at the same time, MPP would still be able to accelerate matter particles moving along the interplanetary disc due to interactions at interfaces between streams, analogous to the way that wind (air) can cause the waters of a shallow lake to circulate.

a. Suppose that matter (ions and electrons) is always being accelerated forward and to the sides by passing MPPs.
b. All objects being accelerated travel along paths of least resistance.
c. So matter being accelerated forward will travel along paths of least resistance.
d. Suppose that the presence of matter/antimatter ahead of the solar wind always ripples space-time in the area into which the solar wind is directed.
e. Any affect change in space-time is always felt over very large distances.
f. There will always be felt paths of least resistance at very large distances.
g. Therefore, the matter flow will always alter course to take the route with the least resistance.
h. Planets have a large effect on space-time because of their large mass.
i. Therefore, large amounts of matter within the solar wind will always be diverted by the ripples in space-time to pass by a planet whenever possible.
j. Also, therefore, MPPs would be expected to avoid the interplanetary disc.
k. All along and through the immiscible boundary, marking areas occupied by matter versus areas occupied by antimatter, field forces can be felt.
l. Therefore, MPPs will continue to accelerate the matter portion of solar wind, forward and to the sides.
m. Therefore, the matter portion of the solar wind will always be a unidirectional flow radiating away from the Sun tending to skip from planet to planet (Conclusion 32).

The boundaries between MPPs and matter-rich areas would be expected to be fluid, transient, and essentially void of either material. A good analogy for what these interactions might look like from a macro perspective would be the action of a sheepdog herding, running alongside and through a flock of sheep. The sheepdog represents the MPP and the sheep are the matter being driven faster and faster onward. More correctly, the metaphor should be of a school of sheep-fish being cajoled by packs of sheepdog-fish, which may either run through the slower moving sheep-fish or to the side of the flock. Collisions don't occur in either case, because both the sheep-fish and the sheepdog-fish would be moving at relatively similar speeds. Therefore, mutual avoidance of each other would be easy.

Detractors might question why we see no annihilation events at the boundaries of the solar wind, since both electrons and positrons are expected under Dominium predictions. The construct of the Dominium model and resulting dynamics of immiscibility might be ignored in favor of referencing "refuting" laboratory results. For example, results from LEP, where scientists recorded many annihilation events between electrons and positrons, could be used to infer incorrectly that the annihilation of positrons with electrons is a "common" event. Such parallels cannot be made because of the engineering design of that machine. The whole structure of LEP was designed to force packets of electrons and packets of positrons to converge at the same point at the same time using strong electric and magnetic fields. This manmade design makes it virtually impossible for collisions not to occur. One interesting fact from LEP is that the packets of electrons and positrons need to be replenished only once every 24 hours. This is intriguing because it is rather a long time for the particles to persist inside a machine specifically designed to cause collisions. In that 24 hours there would have been upward of $10^{22}$ (10,000,000,000,000,000,000,000) times that the packets of particles went through each other, yet only some of them were annihilated. Although this fact was dismissed by researchers, owing to the extremely small size of both particles and the extremely wide diameter of the beam in comparison, it is worth noting nonetheless.

***Conclusion 33: MPPs will avoid any and all matter obstructions in their paths, including high-speed cosmic rays.***

Some might wonder about the following. If a random high-energy heavy cosmic ray ion were shot perpendicular to the path of a large MPP, why don't annihilation events occur? Again, to answer this query one must assume that gravitational effects are felt at very long distances. If an individual positron within an MPP was traveling along the z-axis away from the Sun; and if a high-speed particle was moving along the x-axis, where the origin is the expected point of collision; and if both particles felt the gravitational repulsion affect of each other at a very large distance away; then the two objects would be expected to veer into opposite y-directions and avoid collision. All other positrons within the MPP would feel similar effects. As a result, the MPP might be expected to divide, like a school of fish, to avoid the obstacle, only to reform on the other side. Therefore, immiscibility would be expected to hold in all normal scenarios the MPPs might encounter passing the Earth.

a. The presence of matter/antimatter ahead of the solar wind always ripples space-time in the area into which the solar wind is directed.
b. Changes in space-time are always felt over very large distances.
c. Then there will always be felt paths of least resistance at very large distances.
d. Every individual schooling positron feels a unique direction for least resistance.
e. All positrons are positively changed.
f. Then, the organization within an individual MPP will always be loose because of common electric repulsion of the positrons from each other.
g. Therefore, all MPPs can divide whenever a matter-based obstacle is in their direct path or attempts to intercept.

***Conclusion 34: Positrons that escape the Sun will eventually exit the Milky Way dominium, while the matter portion will stay within this dominium.***

As the micellular positron packets (MPPs) move further and further off into space, they would be expected to continue to fuse, forming streams. These streams would take paths of least resistance to exit our galaxy, which have probably been in use by escaping antimatter for eons. Immiscibility would hold and no annihilation events would take place. The positron streams from our Sun would continue flowing into deep space en route to the nearest antimatter dominium. The matter that had been picked up and swept along for the ride during the first part of the journey past Earth, would eventually find a home somewhere else in our dominium, because this model does not allow such matter to join the MPP for the entire journey between dominia. It is imaginable that, just like dust or snow carried in the wind, places exist where these particles are more likely to accumulate than others, acting as gravitational traps for the blown matter. At these points of accumulation, the ions and electrons will cool. Electrical self-sorting will occur and the plasma will organize first into the coupled state of gas, and later into liquids, solids, and molecules. Such points of accumulation could be imagined to receive donations of matter from the solar winds of many stars, not just one. If time and rate of accummulations permit, it is imaginable that eventually enough material will accumulate to begin forming a new solar system. This scenario matches and explains observations of nebulae (the reddish portion of the cover art of this book shows part of the Veil Nebula). Again, although the existence of nebulae has been known for a long time, until the Dominium model, their formation was a mystery.

a. Recall Conclusion 26: Some positrons will escape the Sun.
b. Recall Conclusion 33: MPP will avoid any and all matter obstructions in their paths, including high-speed cosmic rays.
c. Therefore, no escaping positrons will be evidenced by an annihilation event throughout its journey.
d. All positrons are being accelerated away from the core of the Sun and toward the neighboring antimatter dominium that exerts the largest net attraction.
e. Therefore, all positrons that escape from the Sun will travel unhindered out of the Milky Way (first half of Conclusion 34).
f. All matter that was accelerated away from the Sun within the solar wind is immiscible with the dominial destination of all the positrons that accelerated it.
g. Therefore, all matter carried away by the solar wind will stay in this galaxy (second half of Conclusion 34).

***Conclusion 35: The solar wind will contain an excess of electrons compared to the total charge of all positive ions as measured over a discrete period of time.***

According to the stated predictions of the Dominium model, the solar wind is expected to have two components: a matter component (which is currently being monitored/explored) and an antimatter component (which, although it is immeasurable because of immiscibility, would be expected to be completely positive in charge). Also, as noted in Conclusion 34, positrons that escape the Sun will eventually exit the Milky Way dominium, while the matter portion will stay within this dominium. If these two assertions are correct, then a possible anomaly seems to arise: Is our Sun continuously gaining electrons and acquiring an increasingly large net negative charge?

If our Sun is becoming increasingly more electrically negative, that will affect the ability of MPPs to accelerate away from the Sun. As the star becomes increasingly more electrically negative, the MPPs would accelerate at increasingly slower rates. At some point, it is possible that the system reaches stasis and MPPs become unable to leave the surface of the Sun. The MPPs would be halted at a point where conditions favor immiscibility and no annihilations occur at the fringes between the stationary MPPs and the surface matter of the Sun. The other option is that the excess electrons will be among the material pushed along by the solar wind. This option would align with three standard assumptions: excess charges congregate on the surface of an object; the material swept along by the solar wind originates from the outermost layers of the star; and electrical opposites attract. If this scenario is true, one would expect WIND and ACE to pick up a slight imbalance in charge, favoring negative (–) charges, in the solar wind. Conducting this observation would be relatively easy: count total charge, for each (+) and (–), over the same period of time, then calculate total charge. It seems extremely odd that, to date, NASA has not published the results of such a simple tally. Although this test does not seem to have been conducted, the Dominium model would assert that the matter stream sampled by the WIND and ACE probes would be slightly negative in charge. This an assertion contrary to held assumptions as evidenced by NASA's public information website, which states, "Of course it has electrons too, balancing out the positive charge of ions and keeping the plasma electrically neutral." The net effect to Helios of a net negative ratio within the matter-based stream would be to keep the star itself relatively neutral inside, because the escaping (+) MPPs would be balanced out by the excess of evicted (–) electrons.

The path predicting that the excess negative charges are picked up and taken out of the Sun's immediate sphere of influence is supported by our elementary understanding of charge interactions. If excess charge congregates on surfaces of objects, and the mass that has the best chance of being pushed away is that which is highest in the atmosphere, then excess electrons would be among the matter portion of the solar wind.

a. Recall Conclusion 26: Some positrons will escape the Sun.
b. All positrons have a positive charge.
c. Then, the Sun will be accumulating a net negative charge (electrons).
d. Charge always accumulates on the surface on an object with an imbalance.
e. Then charge will accumulate on the outermost surface (upper atmosphere) of the Sun.
f. Plasma in the uppermost portions of the solar atmosphere is always the most likely to be accelerated away by the MPP drivers of the solar wind.
g. Therefore, the excess electrons will be among the most likely particles to be driven off in the solar wind.
h. Consider a sampling of all matter-based particles passing by a space-probe in the solar wind.
i. Then (prediction): The solar wind will contain an excess of electrons compared to total charge of all positive ions as measured over a discrete period of time.

### *Conclusion 36: The entire Milky Way dominium is becoming increasingly charged, net negative, while antimatter dominia are becoming increasingly positively charged;excess charge will congregate along dominial borders.*

Not only is our star, Helios, leading to a more negatively charged Milky Way, but every star in our galaxy which produces a solar wind would be expected to do so by generating a net negative charge. Cumulative effects would be in place, leading to a very large continual acquisition of negative charge by our dominium. Conversely, given the assumption that symmetry exists on all levels, antimatter galaxies would be expected to be becoming increasingly electrically positive. Adding to the increasing imbalances of charge would be positrons migrating away from dominia made up of matter and the migration of electrons away from dominia based on antimatter. Loss of positrons from our stars and gain of electrons from neighboring dominia will help to make our dominium even more negative. The increasing imbalance of electric charge between dominia would be expected to rise steadily, because the fusion within stars seems to be an overall steady and continuous process with any given galaxy. Excess electrons would be expected to space themselves out along the outermost boundary (surface), where immiscibility is established between dominia and electrical interactions between dominia would increase over time.

a. Recall Conclusion 34: Positrons that escape the Sun will eventually exit the Milky Way dominium, while the matter portion will stay within this dominium.
b. The proposed scenario is the same for all other stars within the Milky Way.
c. Then, the Milky Way is becoming increasingly net negative.
d. Charge always accumulates on the surface of an object with an imbalance.
e. So excess charge will accumulate on the outer surface of the Milky Way.
f. The outer surface of the Milky Way is the dominial boundary with the adjacent antimatter galaxy.
g. Therefore, excess negative electrical charge produced by all the stars of our galaxy will accumulate at the boundary (outer surface) of our dominium.

***Conclusion 37: Increasing electrical imbalance between dominia will eventually destabilize the entire super-matrix of the Universe and lead to the next Big Bang, the supersymmetric equivalent of the creation of a black-hole.***

The increasing imbalance in charge would cause an increased attraction between unlike dominia due to a continuous increase in opposites-attract electrical relationships, thereby slowing the rate of increase of expansion of the entire Universe.

The slowing of the rate of expansion of the super-matrix of galaxies due to this increasing electrical attraction between dominia would not go on forever. At some point in time, the two forces would be expected to balance out between dominia: gravitational repulsion versus electrical attraction. If gravitational repulsion were relatively fixed, compared to the ongoing process of charge imbalance, the system would eventually tip in favor of electrical attraction. As the scale tips more and more in favor of electrical attraction, a force imbalance would be achieved. At this point the Universe might be expected to either stop growing or contract slightly, as the system becomes supersaturated in favor of the electric attraction force. Once supersaturation is established, then any small event could and would lead to a cascade/collapse. When the cascade happens, one would see a cataclysmic collapse of all dominia into a central point—a supersymmetric equivalent to the formation of a black-hole—yet in this case the result would be a conglomeration of all mass/energy and matter/antimatter of the Universe into one spot, which would not be able to last very long—the Big Bang. There is supersymmetric irony in the last statement. Black-holes are the most stable species known to man, while the Big Bang fireball is the most unstable event ever imagined.

a. Recall Conclusion 36: The entire Milky Way dominium is becoming increasingly charged, net negative, while antimatter dominia are becoming increasingly positively charged; the excess charge will congregate along dominial borders.
b. The gravitational repulsion forces are all changing at a slower rate than the electrical attraction forces that are pulling the Universe inward.
c. Then the electric attraction forces will always catch up to the gravitation rate as long as premise "b" holds true.
d. The net electrical force points toward the center of the Universe.
e. The net gravitational repulsion force points away from the center of the Universe.
f. Then eventually the net force will always end up favoring inward attraction, as long as premise "b" holds true.
g. Recall Premise 4: All black-holes form as the result of an imbalance of forces: Gravitational effects favor collapse; other forces act to prevent collapse.
h. And recall Premise 2: All phenomena that occur in electrical systems are directly applicable as a predictor of gravitational behavior, if the conditions are tightly analogous, because of supersymmetry. (But here we are using this premise in reverse—using a known gravitationally driven event [formation of a black-hole] as a predictor for an electrically driven system [Big Bang]).
i. Therefore, the Big Bang is the supersymmetric equivalent of the formation of a black-hole (Conclusion 37).

After that point in time the process would be expected to begin again, and repeat, again and again, infinitely.

The Dominium model provides an explanation for virtually everything we see today. Instead of undermining the Grand Unifying Theory and Standard Model, the Dominium model presents undeniable support for the basic concepts inherent in those constructs—parity and supersymmetry.

### *Conclusion 38: Einstein made at least one mistake.*

The Dominium model begins with sound premises and uses tight syllogisms to construct the its model spanning from Big Bang to Big Bang. Anomalies that could not be explained given conventional models all have purpose and predictability under the Dominium. The Dominium model goes many steps further by showing supersymmetry between gravitational and electric forces, as displayed on many levels. One way to prove Einstein incorrect is to compare the correctness of the tie between the Dominium model's complete descriptions versus the limited accounts and anomalies embodied in current theories. The ability to predict the exact conditions and match observed/known phenomena is the true test of any theory. As a syllogism demonstrates:

a. Suppose the Dominium model predicts all known major cosmological phenomena.
b. All Einstein-based theories predict only some of the known major cosmological phenomena.
c. Therefore, the Dominium model is "more correct" than Einstein-based models.
d. Suppose some of Einstein's equations were correct.
e. Then at least one of Einstein's equations was incorrect.

### *Conclusion 39: Einstein probably misjudged entropy.*

The big question is, Where did Einstein make his mistake? In Einstein's defense, I would argue that for him to make a miscalculation is perfectly forgivable. After all, many discoveries and observations have been made since he died that he could not have employed to perfect his models.

a. Einstein employed all the available knowledge of his day in fashioning his formulae.
b. Since Einstein died, more knowledge has accumulated.
c. Therefore, Einstein did not employ all principles that exist.
d. Nanotechnology is a new field that catalogs, utilizes, and depends on an inherent function of heterogeneous electrically dissimilar species that in repeated trials systems would self-assemble into more orderly designs, e.g., ionic crystalline structures, micelles, microtubules, etc.
e. Einstein did not employ the concept of self-assembly to any of his attempts to define a model for the initial creation of the Universe.
f. Recall Conclusion 5: Self-assembly is a supersymmetric equivalent.
g. And recall Conclusion 7: The current structure of the super-matrix of galaxies and the crystalline matrix of a common stone are supersymmetric equivalents.
h. Therefore, self-assembly was critically important in achieving the current design.
i. Therefore, all models attempting to describe initial creation need to account for/include the advent of self-assembly.
j. Suppose "e" above is true.
k. Therefore, the absence of an account for self-assembly during the initial moments of the Big Bang is a sizable problem within Einstein's construct.

Again, it should be expected that Einstein's equations did not include discoveries that happened after his time. For instance, particle physics was in its infancy, experiments were just beginning, and the line between science fact and science-fantasy was still blurry. The existence of a supermassive black-hole at the center of the galaxy was not known. The very question of black-holes hadn't even been asked or answered. The mass distribution of the sky hadn't yet been calculated and imaged. The existence of "dark matter" hadn't yet been detected. The phenomenon of self-assembly had not yet been observed and recorded. Dynamics on the atomic, molecular, or nano scales were not as fully understood as they are today (See discussion: "Dinosaur problems.") Without these observations and experimental results, the line of inquiry used by this Dominium model would be impossible to justify. But given these modern discoveries and experimental results, the Dominium model is hard to refute.

***Conclusion 40: The Tunguska event was a near-Earth encounter with an MBH.***

Given the acceptance of the Dominium model's prediction for the formation of micelle-black-holes (what have been measured as "dark matter") and given that those micelle-black-holes with asymmetric geometries are expected eventually to exit our dominium, the following construct is a tour of mechanisms, repercuss-sions, and plausible evidence of interactions between the Earth and a passing antimatter micelle-black-hole.

It is generally assumed by the Dominium model that because of the repulsion forces between opposites, migrating micelle-black-holes will avoid our little corner of the Universe. This would be especially true for extremely large micelle-black-holes, which would be more affected at a further distance by gravitational effects of the Sun, Jupiter, and Saturn, so as to avoid this solar system entirely. However, smaller micelle-black-holes could, conceivably, squeeze through the gravitation eddies of the solar system's three largest members and head toward Earth.

Depending on its speed and initial trajectory, a micelle-black-hole heading toward Earth might be expected to bypass us at a distance. Again, this would be expected to occur with the medium- to large-sized micelle-black-holes, because they would experience a greater cumulative force imparted on them. Again, it is conceivable that a micelle-black-hole could be small enough and true enough to its course that it could come quite close to Earth, or even enter the atmosphere.

If a antimatter micelle-black-hole did enter the atmosphere, it would not be expected to produce an annihilation event. Rather, immiscibility would prevail as a central assumption of the Dominium construct. Therefore, if it did enter the atmosphere, an envelope of void space would encapsulate and insulate the micelle-black-hole from the possibility of contact with any Earth-matter. The velocity of the micelle-black-hole would continuously slow as the micelle-black-hole approached the Earth's surface. The shape of the path of the micelle-black-hole would be expected to appear as an inverted parabola—a reversed curve of the normal drawing of projectile motion. Effects to the surface of the planet would depend on what was immediately below the negative apex of the curve. The net effect would be as if everything were being pushed: trees would be flattened, buildings would be knocked over, the ocean would slosh into tsunami. There is a record of such an occurrence. In Siberia near Tunguska a large swath of forest was toppled[Tunguska] in 1906. All the trees seemed to have been broken in the same direction. Fortunately it was virtually uninhabited land so no one was killed, but that meant that there were no good witnesses to the event either. Though it was long assumed to be the result of a meteorite, no traces of such a meteor were ever found, and neither is everyone within the scientific community satisfied that a meteor could have created the pattern of damage that was evidenced. Entry of a small micelle-black-hole into the atmosphere above this position would be an explanation for this event and the pattern of damage imparted.

a. If an MBH entered Earth's proximity field, repulsion forces would always be imparted between the two masses.
b. If an MBH entered Earth's atmosphere, its trajectory would always assume an inverted parabolic path and employ projectile motion principles.
c. If an MBH came close to Earth it would exert gravitational repulsion field forces that could be quite powerful.
d. Therefore, the force would be distributed on Earth in a manner reflective of the parabolic path of the projectile MBH.
e. Therefore, the effect of such a force will always produce a geometric distribution.
f. The effect of immiscible matter interacting with antimatter is to push obstacles forward and to the sides.
g. The Tunguska event of 1906 flattened a forest with an apparent geometry in which the trees appeared to be toppled forward and to the sides.
h. Therefore, the Tunguska event was a plausible near-Earth encounter with an MBH.

Therefore, the Tunguska event gives evidence of the possibility of the existence of very small micelle-black-holes. As mentioned above, large- and medium-sized micelle-black-holes would feel the effects of Helios, Jupiter, and Saturn long before getting close to the Earth.

Again, this evidence of the possibility of *small* black-holes is of the utmost concern to current operations at CERN. LHC, Brookhaven, and other labs around the world are currently in a race to be the first to create a small black-hole. Scientists and proponents of LHC have given multiple assurances that a small black-hole will not be stable because the collapse of supernovae into black-holes only occurs among the largest of stars. The Dominium model has now shown in multiple ways that such assurances are based on faulty assumptions.

***Conclusion 41: The existence of small micelle-black-holes is predicted by the Dominium model.***

This conclusion is based on earlier conclusions 9 through 12 and 40. When the Big Bang fireball expanded enough to dilute energy per regional volume, the conditions for immiscibility would have been established. Almost concurrent with the creation of immiscible boundaries that separated regions was the Dark Event, where micelles toward the center of every dominium were compacted into micellular black-holes (MBHs). Proximity within the embryonic dominium was identified as the main determinate for which micelles were most likely to experience inward forces strong enough to overcome any outwardly directed electrostatic stabilizing force. Micelles that were most likely to be compacted were those toward the center of any given dominium. Since micelles toward the center of the embryonic dominium would have varied in size, the MBHs produced in this event would have been all different sizes.

*a.* Recall Conclusion 8: Micelles of opposite-type material existed in embryonic dominia/galaxies.
*b.* Micelles present would have varied greatly in size.
*c.* And recall Conclusion 9: Soon after the advent of immiscibility, micelles of underrepresented antimatter collapsed into micelle black-holes.
*d.* Then micelles of all sizes were collapsed during the Dark Event.
*e.* Suppose that small micelles were collapsed in all dominia.
*f.* Then, small MBHs were formed in all dominia.
*g.* Suppose that symmetry exists between the behavior of matter and that of anti-matter.
*h.* Then a small matter-based black-hole could exist.

The sequence of events that could have led to the Tunguska event:

1. Soon after the Big Bang, conditions were met to achieve immiscibility.
2. The advent of immiscibility happened during the micelle-rich stage, and the micelles of all sizes were collapsed into micelle-black-holes.
3. Micelle-black-holes, being repulsed by the dominia in which they were created, either escape or get trapped as a result of regional geometries.
4. Micelle-black-holes in transit to the closest like-type dominium are what has been picked up by the SDSS and labeled as "dark matter."
5. A small micelle-black-hole, late in escaping this galaxy, is what caused the Tunguska event.

The Tunguska event would be expected to be a highly rare occurrence since most of the small micelle-black-holes that were produced in this dominium would have escaped eons ago. Another reason to conclude that the Tunguska micelle-black-hole was "small" is the fact that the area of woodland affected was relatively small compared to the face of the Earth.

***Conclusion 42: LHC is a machine that could create a matter-based black-hole on Earth in the not too distant future.***

LHC is a marvel of technology. When it is turned on, it will enter the *Guinness Book of World Records* as the largest machine ever built. Its construction was a model of international cooperation, employing people all around the world. Its design consists of two antiparallel beams of protons traveling in opposite directions that are forced to cross paths at manmade junctions. Surrounding the junctions are the finest examples of detector and computing technology installed to monitor and record events occurring at those junctions. Many hope that through the much higher energies that will be established within, this machine will provide evidence for the supersymmetric particle hypothesized to exist (known as the Higgs' boson). Many openly speculate about the chances of creating man's first synthetic black-hole. The reason for this "hope" is that protons are going to be piling up in one place at extremely close proximities. If gravitational attraction does balloon at extremely close ranges, then there is a high likelihood that protons will be forced so close to each other that they will experience ballooning effects. Past experiments have attempted the creation of black-hole material; however, only a tiny conglomeration of protons was ever achieved and it degraded quickly into normal forms of Earth-matter. Beams within LHC will contain packets containing upwards of $10^{13}$ individual protons. Junctions will direct these packets of protons to collide and pile up. Once a pile-up begins, subsequent packets will collide with the first. Very quickly, a snowballing of increasingly larger amounts of material tightly packed within one very small space will occur—this is the scientists' hope. If large amounts of protons are piled up at one specific junction, as in the design of LHC, then much larger conglomerations will be achieved. Therefore, with a large enough amount of matter in one place, as may be possible with LHC, the threshold may be exceeded and a stable black-hole species be created.

a. LHC is designed to force large amounts ($>10^{14}$) of protons to collide into similarly large amounts of protons.
b. Recall Premise 3: Gravitational forces will experience exponential increases in effect at very close ranges.
c. Therefore, large amounts of protons will be in close enough proximity to feel ballooned like-like gravitational attraction.
d. Recall Premise 4: All black-holes form as the result of an imbalance of forces: Gravitational effects favor collapse; other forces act to prevent collapse.
e. Therefore, the ballooned gravitational forces may be large enough to result in a stable, small, synthetic, matter-based black-hole.
f. LHC is a machine that will attempt to force large amounts of protons to come in close proximity with other large amounts of protons.
g. Therefore, LHC is a machine that could create a matter-based black-hole here on Earth in the not too distant future.

This now becomes a question of risk and a question of ethics. Yes, the exploration of science is important. The exploration of the unknown has always involved risk. For centuries, scientists have willing laid down their lives for their research. The difference is that today's risk involves the entire planet, not just those in the lab.

***Conclusion 43: A moratorium needs to be set on any further attempts to create a synthetic matter-based black-hole.***

Once the black-hole is created, assuming that the diligence of the scientists is successful, what then? There are several important questions regarding the consequences of the creation a black-hole once it exists in the lab: How does one reverse the process or dispose of a sample that was created; how does one contain and store a black-hole, once a sample quantity has been produced; and how will a stable matter-based black-hole behave in the matter-rich environment of the Earth? Currently it is impossible to know with 100% confidence the answers to any of these questions. There are those who profess to have equation-based models that make predictions and projections. However, any such model based on any of Einstein's flawed assumptions would be expected to yield incorrect results, given the results of the Dominium model. Of course, scientists and spokespeople for LHC will assure the public that all aspects of LHC have been thoroughly planned and considered, and that therefore, there is no need to worry. These people are vested in this project through many years of anticipation and planning. Yet, even the most decorated scientist can only speculate, because there is no physical evidence or test result from the past that could fashion a conclusive response to any of the three questions listed. In this discussion, the three unanswered questions regarding man-made black-holes will be explored. Not even the Dominium model is equipped to give categorical answers to these important questions; however, the model is a useful predictor.

*How does one reverse the process or dispose of sample of stable black-hole material once it's synthesized?*

This question contains a degree of paradox. Essentially, the question is: How do you destabilize the material once you've stabilized it? If the sample is stable, then it's stable ... it won't be popping back into Earth-friendly material by itself. Another important related and unanswered question: How will we know when our synthesized black-hole material is nearing stability?

Black-holes are best known for their ability to feed. Many observations have been made of black-holes consuming large quantities of mass in very short periods of time. However, no observations exist of the reverse process; once matter is consumed it stays consumed. True, sometimes black-holes have been noted to "burp" when consuming extremely large quantities of mass at one time. Such observations indicate sloppy eating habits, not instability. Black-hole formation is considered by most to be a terminal step in the lifecycle of giant stars within the universe—there is no known next "more stable" version. Change is seen as a constant dynamic within most systems, one stage progressing on to the next. Black-hole formation seems to be an exception to that rule. Once matter enters a black-hole, it appears to stay inside the black-hole. There is no next step that this system will evolve into, other than, with each bit of mass consumed, the black-hole becomes more able to attract and consume more mass. Other than continuous feeding, there does not appear to be a lifecycle for black-holes. There have been no recorded destabilizations of a natural black-hole becoming something else. Therefore, if a stable matter-based black-hole is synthesized in the lab, it would be highly unlikely that it could be destabilized into non-black-hole material again, as the following syllogism shows:

a. All observed matter-based black-holes appear to be stable and to continuously feed within our matter-based galaxy.
b. No evidence for a destabilization event of a black-hole has ever been recorded.
c. Therefore, all black-holes will remain stable and continuously feed when situated within an area of like type material.
d. Recall Conclusion 42: LHC is a machine that could create a matter-based black-hole here on Earth in the not too distant future.
e. Therefore, any stable black-holes created by LHC would remain stable and continuously feed (supporting premise 43a).

*How would one contain a stable synthetic black-hole once it's been made?*
Once black-hole material is made, it would be expected to behave differently than its ingredients had before. There is near certainty for this conclusion, because the black-hole species is a completely different state of matter. The difficulties experienced in containment will be similar to those during early trials for the creation of anti-hydrogen. To conduct that experiment, researchers were required to take two charged anti-particles (a positron and an antiproton) and combine them to make a neutral atom. The electric fields that were so useful in setting the stage for the creation of the new anti-atom suddenly become useless once you've made your sample. All of the early experiments toward the creation of antihydrogen ended with the neutral anti-atom crashing into one of the walls of the containment instrument. According to the Dominium model, annihilation (and therefore miscibility) occurred because the rigid tightly dense electrostatically bound solid wall enacted the conditions for miscibility. Judging from experience, early experiments attempting to fashion a synthetic black-hole will be exactly the same: once it is made, its containment requirements will not be perfectly met and it will crash into the wall of the instrumentation that surrounds/records the event. Looking back through the histories of other disciplines of science, flawed initial containment and handling is the norm. Whenever one tries to handle a new species, the initial experiments will be imperfect in terms of containment and handling because of ignorance. Consider for example the Curies with radioactive elements; the United States, USSR, and China with respect to aboveground nuclear testing; or CERN's antihydrogen project. All of these examples represent periods of great advancement in scientific understanding. Tragically, all three of these revolutions were marred with problems regarding initial imperfect handling and containment of samples generated. There is no fault in pure ignorance. Although all members involved in all three examples contributed greatly to science, none of them knew for sure what would happen next when they conducted their experiments. Therefore, a safe premise is that all early experiments to contain and/or handle an unknown species are inherently imperfect. With that in mind, the following syllogism results:

a. Recall Conclusion 42: LHC is a machine that could create a matter-based black-hole here on Earth in the not too distant future.
b. All early experiments to contain/handle an unknown species are imperfect.
c. Therefore, LHCs early attempt to contain/handle the unknown species, matter-based black-hole material, will be imperfect.
d. Therefore, samples of the first synthetic black-hole material made at LHC will collide with a wall of the surrounding container or instrumentation (supporting premise 43b).

*How will matter-based black-hole material behave in the matter-rich environment of the Earth?*

Good question. No scientific work is intended from the start to be alarmist; but the Dominium predictions are truly alarming. The annihilation of antihydrogen by colliding against a wall during early experiments is a good predictor of what might happen to a stable sample of black-hole material. In LHC, as it was with the antihydrogen project, initial containment won't be 100%. Also, both experiments used inertial beams of particles to create their new species; therefore, LHC black-hole samples would be expected to exhibit inertia after the creation is complete. Some speculate that black-holes will be safe to handle on Earth because the "event horizon" of a tiny black-hole would be so short that it wouldn't interact with anything. Such musings go against what has been established about black-holes: they continuously feed. If a black-hole grain smashes into the wall of the instrumentation, such a collision would force the new synthetic black-hole into the wall of LHC. This collision would force the new black-hole to interact with normal matter—essentially force feeding the tiny Beast. With every speck of matter that it incorporates into itself, it would become more and more easy for it to feed itself. Whether it would move unhindered in parabolic motion as it fed or would tunnel and burrow as it fed is trivial. It would feed. It would feed and feed. Eventually, it would grow and get heavy. Gravitational attraction would then cause it to sink to the core of the Earth. Once situated it would feed continuously. The gravitational pull that all things on Earth feel would cause matter to be drawn down to the feeding man-made black-hole. The black-hole would feed until there would be nothing left for it to consume. The entire process is more likely to be quick rather than slow, since we've recorded celestial black-holes consuming large quantities of matter in very short periods. On the people-inhabited crust of the Earth, the scene would be apocalyptic. Ironically, fire and brimstone would both be expected as the crust heaves and cracks while the volume underneath it implodes. Other areas would sink only to be swamped and devastated by floods as the oceans rush in. Although the exact time frame is unknown, the certainty of these eventualities is clear: the process would continue until all is consumed. As the black-hole feeds, mass that once occupied substantial volume will be reduced to mere nothingness. The Earth will lose internal volume and implode upon itself. Earthquakes will be massive as the planet continually adjusts to its shifting internal volumes. Liquids will be drawn to the lowest spots of the imploding Earth, causing floods. As the volume of the sphere of the Earth decreases, its surface area won't be as large as it is today. To achieve its new volume-dependent surface area, the crust would buckle and fold. As this process occurs, mega rift-volcanoes will spew toxic gases, molten materials, and solid chunks. Toxic gases will spew from these massive rift-volcanoes and choke the air.

Whether or not the newborn Beast would cause visible havoc at CERN itself is debatable and uncertain. True, one possibility could be a new volcano emerging at that cite from the borehole of the feeding matter-based mini black-hole. Alternatively, another scenario exists where the apocalyptic monster is created with very little fanfare other than breaking the vacuum seal on the LHC machine after it is conceived. When it is first created, the little Beast will be very little, smaller than the leg of a mosquito. Because of its size, its ability to feed will be small, but feed it will. As a result, its initial damage will also be on the microscopic level. The hole it bores through the hull of the LHC machine will be barely noticeable, but be large enough to allow ambient gases to flow in and destroy the manmade vacuum environment. Although very small, it will feel the tug of gravity and float down through the earth as described above. However, the borehole that it leaves will also be very small. Therefore, lava will not be able to flow through it. All the same, it will be continually growing in mass and therefore growing in its ability to feed. Once it reaches the core of the Earth, it will be force fed by favorable inward facing gravity vectors. Again, hour glass affects will be in place and the Earth will slowing implode upon itself. From this point on, the description will be the same as in the preceding paragraph: earthquakes larger than ever experienced before, tsunamis, lava, sulfurous gases, brimstone, etc.

One of the most chilling aspects of the predicted sequence of events is the parallels to end-of-the-world narratives spelled out by the Bible and the Quran. Fire, brimstone (sulfur), and floods are all predicted by the scientific Dominium model, just as they are by scripture. The idea of everything on Earth being sucked down toward the core parallels the notion of being dragged down to Hell. In Islam, it is said that some will take as long as 80 years to be dragged down to Hell; this concept is compatible with the Dominium's doomsday scenario. Whether or not you believe in either of these organized religions, the similarity between the narratives is striking.

There are those who will try to comfort themselves with the thought, "At least the whole process will be quick and painless; after all, we've observed whole worlds swallowed up by a black-hole in a micro-second—I won't feel a thing." Unfortunately, such people are making a false comparison of the dynamics of a giant black-hole with the effects of a mini-black-hole. Giant black-holes that have been observed are composed of masses that are thousands, millions, or even trillions times bigger than our Sun. Resulting from that enormous size would be a huge gravitational pull, and a great ability to draw in and crush objects. A synthetic black-hole, on the other hand, would start off puny; its gravitational attraction and ability to feed would also be very small. However, like all stable black-holes, it would do one thing: consume anything that it can. Therefore, the Quranic projection that the process would take 80 years before the last living person is extinguished could easily be correct. Pity such a person: they would be like a termite trapped on a log in the middle of a just-lit fire—the final fate would be set, but they would try to dance away from the flames, all the same. Because of the very slow process initially, it is likely that no one will be taken "unaware" of the coming doom. Just like in the colony of termites trapped in/on a log of a fire that has just begun, word will travel fast. Despite the scrambling and jostling that ensues, every last worker, soldier, king, and queen will be slowly (at first), but surely, taken out one-by-one.

Given the rate at which LHC is currently being readied, the little Beast could be conceived either this year or next. Given also the expectation that it will take time for the process to be completed and time for the little Beast to gain enough mass to begin feeding rapidly, it could easily be two to six years before apocalyptic conditions are experienced on the Earth's surface. Ominously this timetable could very easily match with the ending of the Mayan calendar in 2012. Is this ironic omen a certainty? If it were, I would not be typing these words. In the deepest realm of my heart I truly believe that we are still at a crossroads up until the very last moment that the little Beast is born. I realize that to make the metaphysical observations of ironic coincidences between Holy Scriptures and the Mayan calendar imply a belief in God, which is an anathema to the fiercely atheistic particle physics community. For that I am unapologetic. It also implies the validity of my on personal beliefs. But if it is true, then this book could be the final warning to us all. (See Conclusions 45 and 46)

Very few, if any, people would live long enough to actually feel the black-hole compressing their body down to nothingness. Most everyone will be burnt, crushed, suffocated, or drowned while the Earth collapses in on itself. Eventually, nothing would be left except the satellites still fixed in their orbits above black space where the Earth once was. The moon would be "unaffected," continuing the dance it plays out in its orbit around the center of gravity of the Earth-black-hole, though it would seem lonely. The last humans left alive would be the astronauts still in the space station, who would most likely starve to death.

a. Supporting Premise 43a: Any stable black-hole created by LHC would remain stable and would continuously feed.
b. Supporting Premise 43b: Samples of the first synthetic black-hole material made at LHC will hit a wall of the surrounding instrumentation.
c. Therefore, any stable black-hole at LHC will hit a wall, interact with matter, and continuously feed.
d. All black-holes continuously feed in the presence of new material to incorporate.
e. Then, the synthetic black-hole at LHC will feed until it consumes all material on Earth that it can consume.
f. All material on Earth could be consumed.
g. Therefore, the synthetic black-hole will not stop feeding until all the material of Earth is consumed.

As a human, I want to stop here and pose one normative question: When is the risk too high for scientific exploration? Risk-takers are cataloged and glorified throughout the history of science. If people didn't take risks, nothing would ever have been discovered. Braving jungles, taunting sharks, traversing wastelands, colliding particles, mixing potions, and trying "new things"—that is the human experience of scientific exploration. However, there is a big difference between past examples of risky behavior leading to fantastic discoveries and the risky behavior of trying to make black-hole material at LHC. In the past, it was just the wild explorer and/or experimenter and those immediately in proximity who were put at risk of an experiment gone wrong. The danger and ramifications of the "successful" creation of stable matter-based black-hole material at LHC, Brookhaven, or elsewhere on Earth, puts the future of all known life forms in jeopardy. Therefore, a moratorium needs to be agreed to by the world's political and scientific communities against further attempts to build matter-based black-holes on Earth.

***Conclusion 44—There is a way to save both the Earth and the LHC project.***

Although it is an established conclusion of this paper that a moratorium needs to be set against further experiments that could produce a matter-based black-hole, the creation of an antimatter-based black-hole is not prohibited by Dominium directives. Antimatter black-holes do have destructive potential, as evidenced by the Tunguska event (Conclusion 40). However, the detrimental effects of an antimatter-based black-hole experiment gone bad would be limited to the laboratory of its creation. If symmetry truly does exist between matter and antimatter, then the behavior of all antimatter-based black-holes synthesized in the lab could be used to extrapolate the expected behavior of the more dangerous matter-based type. To make an antimatter black-hole, one need only conduct the same experiments, as designed for LHC, Brookhaven, and elsewhere, substituting antimatter in place of matter. For LHC, that would mean using antiprotons instead of protons. True, protons are very easily produced (just uncouple hydrogen), whereas producing antihydrogen is more involved, requiring an initial collision with enough energy for a pair production yielding an antiproton followed by a subsequent purification of the sample. However, in recent years methods for producing and storing antiprotons have been developed and are continuously being perfected by labs around the world, like CERN's LEAR and antihydrogen projects. Only "minor" adjustments will need to be made to LHC to make it compatible with accelerating antiprotons. LHC was designed with the charge, momentum, and behavior of protons in mind. Antiprotons and protons are extremely similar: they have the same absolute magnitude of charge; they have the same mass and momentum; and they respond similarly to stimuli. Therefore, LHC should be as functional with antiprotons as it would have been with protons.

a. All aspects of LHC were designed with protons in mind.
b. All fundamental characteristics of both protons and antiprotons are exactly similar in terms of magnitude and behavior.
c. Then LHC was designed with all aspects that would be important for the study of antiprotons.
d. Recall Conclusion 42: LHC is a machine that could create a matter-based black-hole here on Earth in the not too distant future.

e. Therefore, LHC is a machine that could create an antimatter-based black-hole on Earth in the not too distant future.
f. Understanding the dynamics of antimatter black-holes is an area of science that has never been traversed.
g. Then it remains legitimate territory for exploration.
h. All the Earth could be consumed by a matter-based black-hole.
i. The worst damage ever caused by an antimatter-based black-hole in recorded history was the devastation of the Tunguska forest.
j. Therefore, an antimatter black-hole is less of a potential threat to the planet.
k. The dynamics of both types of black-hole are unknown to science and worthy of discovery.
l. Therefore, LHC should change course and proceed to make antimatter black-holes, while experiments to create a matter-based black-hole should wait until they can be conducted some place other than Earth or on an Earth-satellite with no potential of ever falling to Earth.

This set of theories needs to be read by the "right" people—or enough people—who might be able to divert the human momentum building up for the start-up of LHC and the potential creation of a stable matter-based man-made black-hole. I hope truly there is a mistake in the construct leading up to Conclusions 42 through 44; however, all inferences used appear to be valid, and the succession of conclusions sound. I am not "happy" about the direction in which the Dominium model points, yet I cannot deny the validity and soundness of the construct.

***Conclusion 45: The existence of God is not addressed by the Dominium model; the consensus of major religions states that God permeates throughout time and space, and is always accessible.***

Although some scientists would disagree, no model delineating the aspects of the absolute creation and continuance of the Universe would be complete without mention of God. After all, the notion that some force/entity was responsible for the creation of all is found in almost all cultures on Earth. To ignore the question and place of God would be "unscientific" in its incompleteness. Within every religion there are people convinced that they have, at least once in their lives, experienced a connection to that invisible motivator. Who's to say that such people are delusional? On the contrary, because of the commonality of spiritual feelings among mankind, the opposite conclusion should be made: Perhaps, those moments of true connection that some people report feeling is real. Just because "I" haven't recognized it, doesn't mean that it wasn't there all along. Examples of such realizations are commonplace. Therefore, an analysis of God is a necessary task of this work.

Unfortunately, the Dominium model itself does not offer any suggestion either for or against the existence of God. Therefore, this particular analysis is disconnected from all other prior work on this model. The approach will be similarly based in Aristotelian principles. Therefore, the first task of this analysis will be to find initial categorical premise(s) that hold true for all religions. Although on first glance, the great religions of the world seem very dissimilar, a deconstructionist approach does yield a common denominator. In no particular order:

· The Catholics say, "God is omniscient, omnipotent and omnipresent."
· The Protestants say, "The Lord is my shepherd."
· The Jews say, "God is eternal."
· The Hindu say, "Shiva is ever watching."
· The Buddhists say, "The Path is ever before you."
· The Moslems say, "God is the biggest."

Therefore, if this one premise is accepted, one can conclude that God permeates time and space. If that one central strand is correct, then it doesn't matter which way you search; as long as your heart truly searches to find him/it/her/us, you will find him/it/her/us. All directions will lead to aspects of absolute truth and beauty, as defined by his/its/her/our presence, because with this one premise one must conclude that he/it/she/we lie in all directions. If something lies in all directions, then in terms of science, one can conclude that that thing—God—permeates time, space, and our lives. All paths to God are, therefore, valid; it's the heart of the individual that must hold pure to course. Not only that, inherent in the doctrines listed above is the idea that God is always accessible to those who haven't yet started searching.

***Conclusion 46: God intends for us to protect the Earth, e.g., divert projects that could extinguish all life.***

An analysis of world religions yields a second common denominator in terms of humanity's responsibilities to the planet on which we live. Scriptures describe man's role as "steward," "guardian," "master," and "caretaker" of this planet. Within that list of descriptions, the idea of being the destroyer finds no place. There is an enforced law on the books in the Commonwealth of Massachusetts that makes it unlawful for anyone to abuse, willfully neglect, or cause damage to any structure or residence making it uninhabitable, even if you are the full owner—no liens or mortgages. The world community, at the height of the Cold War, banned aboveground nuclear testing, in light of subsequent statistically significant increases in the rates of infant leukemia in populations downwind of test sites. I know personally about this historic event: my would-be sister died at two years old, one year before I was born, while my family lived in Pennsylvania—down wind from the Nevada and New Mexico test sites. I remember knowing several other children who had leukemia in my church school classes growing up. Eventually, my parents moved the family upwind to California, partly because they didn't want to bury another child. Meanwhile during that same period of history, the USA, China, and USSR did meet—even when diplomatic relations were extremely strained by the Cold War—and plume-pattern statistical increases in incidence of infant leukemia were shown downwind of all test sites in each country. A moratorium was set and aboveground nuclear testing ceased. The world community is therefore capable of changing direction when human welfare is at stake. I hope that still holds true, but this time action needs to be preemptive.

The strangest aspect of this work is the dominion's role within this new model. Although I am the "author" of this work, I cannot take credit for most of what's written. The truths spelled out in the preceding syllogisms and conclusions came to me in a flood of revelations. I don't like using that last word of the preceding sentence, but there is no other way to describe what I've experienced. At times I felt like I was transcribing, rather than writing or thinking. A net result is that my hitherto dormant love of God has been reawakened. Rather than feeling proud and accomplished as a result of these writings, I feel small, insignificant, and too weak to accomplish the task that seems to be put before me. From the very beginning I felt a sense of doom, dread, and urgency in my role of getting these writings completed and dispersed to the world community. This dark feeling preceded all revelations concerning LHC. It wasn't until I was nearly through transcribing the Dominium construct that manifestation of the dangers inherent in LHC's current direction became clear. I don't mean to presuppose the will of God, but if the existence of the entire planet is in peril, and if God truly loves all creatures, then this would be a time when intervention would be expected.

I am just a simple man living a simple life as an inner-city veteran high school science teacher. I have nothing monetarily to win or lose from the set-up of LHC; I hadn't even thought about that project for almost seven years. I don't belong to any sect, church, or other religious group, yet I have always believed in God and tried to live my life accordingly. I don't "want" the chaos and upheaval that release of these writings might inflict on my life. Yet, there appears to be no other path for me to take. I have a conflicted sense of what the future will bring: one option seems to lead down a path filled with great joy and reconciliation for all mankind, while the other path seems to lead to pure doom and destruction. Man's ability to make choices is what sets him apart from all other lower life-forms. It appears that now is a time to make a choice.

### *Conclusion 47: Hierarchy within Supersymmetry of Gravitational and Electric Phenomena as Explained through the Dominium Model.*

Supersymmetry is the idea that two forces within nature can have a deep symmetry with one another. There are nine distinct places within the Dominium model that hint at supersymmetry, and one anecdotal example known to every first-year algebra-based physics student that indicates supersymmetry. Also, there is the fact that using Premise 5 (supersymmetry between electric and gravitational forces) ultimately leads to a complete and clean model.

As was explained in the first part of this model, the notion of supersymmetry (Premise 3) asserts that gravitational forces will experience a ballooned force effect at very close ranges. In electrical systems it has been measured that electric forces balloon from force being proportional to one over distance squared ($F \propto d^{-2}$)

jumping to force being proportional to one over distance to the seventh ($F \propto d^{-7}$). Assuming a supersymmetric relationship, this premise sets up the conditions that would be expected during the earliest moments of the Big Bang fireball.

Shortly after the creation of the Big Bang fireball, the density of energies would decrease across the entire system because of expansion. These conditions lead to the next use of supersymmetry in the Dominium model: the notion of immiscibility (Conclusion 2c). The idea of immiscibility is the supersymmetric equivalent of the establishment of boundaries in heterogeneous electrical mixtures such as hydrophobic/hydrophilic substances that have been forced to mix through the input of energy. Once the energy responsible for causing the mixture dissipates, immiscible boundaries will form.

Another process that hints of supersymmetry and is connected to the notion of immiscibility is the self-assemblage of a mixture to realign into a more ordered system (Conclusion 5). As proven over and over in the field of nanotechnology, self-assembly is so certain that it is used as a tool of the nanotechnologist to design and build macro-nanostructures. Self-assemblage of dominia would be expected immediately after the heterogeneity of dispersion of matter and antimatter had been established. There is certainty in this prediction; both cases involve opposing forces among mixed-up neighbors—some tugs or pushes will be stronger than others. To compensate for the imbalances in tugs and pushes, all members will shift slightly to achieve stasis. The prediction that the system will adjust to achieve stasis is rooted in very firm foundations found everywhere you look within nature.

The prediction that the Universe would mature into a structure analogous to a common stone (Conclusion 7) is the next supersymmetric analogue. This conclusion comes from observations of heterogeneous mixtures of elements that are allowed to cool over time. The example used for this conclusion is that of magma (a heterogeneous mixture of elements and ions) undergoes self-assembly to achieve a crystalline matrix, as evidenced in the atomic structure of rocks.

The shape of any antimatter trapped within a matter-dominant dominium, such as the embryonic Milky Way, was predicted to be spheroid in Conclusion 8. Again, this conclusion comes from supersymmetric understanding of heterogeneous electrical mixtures, such as hydrophobic/hydrophilic electrically based systems. Probably the most common structure of self-assemblage is the formation of spheroid "micelles." This term is borrowed and employed for the use of this model because of the strong expected supersymmetric parallels in form and function.

The next observation of supersymmetry comes in the explanation of the events that led to the formation of the central culvert black-hole in the centers of known galaxies (Conclusion 19). The event itself is not a supersymmetric equivalent, but the conditions that halted the event from collapsing the entire galaxy do have supersymmetric ties. The notion that the accumulation of micelle-foam around the perimeter of the forming black-hole runs in parallel to multiple examples of stasis events found in electrical systems.

Yet another observation of supersymmetry comes in the prediction of a flat event horizon (Conclusion 23). The supersymmetric equivalent leading to this conclusion is the tertiary structure of sheet-proteins. These molecules achieve their shape because of alternating positioning of charged members, which would be analogous to the alternating placement of dimple-down versus dimple-up mass which occurs after self-assembly or dominia.

The next observation of supersymmetry comes in the discussion of hierarchy of forces (Conclusion 25). The supersymmetric parallel is drawn between the uncoupled plasma-state of electrically driven atoms with the miscible states of high-energy gravitational systems during the Big Bang and within the interior of stars. Concurrent in this discussion is the related supersymmetric parallel of electrically coupled atoms to conditions of immiscibility. Finally, within this same discussion is the supersymmetric understanding that both conditions uncoupled/coupled and miscible/immiscible are energy dependent.

The last major identified example of supersymmetry is the relationship of the creation of a black-hole to the Big Bang. Both events are caused by an ongoing process that eventually tips out of balance. The fact that the processes and forces that eventually tip the balance are mirror images of each other is further evidence of supersymmetry. There is also supersymmetric irony between the formation of black-holes and a Big Bang. Black-holes are the most stable known forms of matter, while a Big Bang fireball is the most unstable condition known in nature.

Besides the nine identified examples of supersymmetry within the Dominium model, there is also anecdotal evidence of this relationship. Students taking algebraic introductory physics across the globe have recognized the anecdotal evidence for supersymmetry between gravity and electrostatic forces within the formulae. If one looks at the structure of Newton's Universal Law of Gravitation and compares that to Coulomb's Law for electrical interactions, they appear very similar. In both laws, a force felt by each of two "things" is proportional to the product of the "things" themselves divided by the squared distance between them (as shown in the following formulae):

$$F = \frac{kq_1q_2}{d^2}$$

Coulomb's Law

$$F = \frac{Gm_1m_2}{d^2}$$

Newton's Universal Law of Gravitation

Many teachers even present these two formulae together, because besides being structurally similar, both describe the effects of a field force. Field forces, unlike contact forces, can be felt at a distance and even through a vacuum. These "anecdotal" similarities thus point to a supersymmetric connection between gravitational and electric forces.

Although the goal of this work was not to prove supersymmetry between gravitational and electric forces, that is an ultimate outcome. The goal of these articles was to explore one tiny possibility...what if "like attracts like and opposites repel" describes a valid scenario for gravitational interactions? The Dominium model was the sequential catalog of ramifications of what this might mean. Like any puzzle, each conclusion led to a new consequence. As the construct "materialized," one unexplainable anomaly after another was predicted as a direct consequence of this model—the expansion curve, the black-hole at the center of the Milky Way, halo structures, unseen but recorded mass, a uniform distribution of the Universe, a flat event horizon, the solar wind, nebulae formation, the cause of the Big Bang, and finally supersymmetry. One mystery that this model does not preclude or address is that of the existence and dominion of God. During the experience of transcribing this model, I have felt especially connected to God. There is no rationale for this. Deep in my heart I feel that this model brings with it a time for all humanity to rejoice. The beauty of God is in nature all around us, super-Supersymmetrically.

However, rejoicing now would be premature. This model shows that small black-holes are a stable phenomenon. This model also asserts that black-holes are the most stable form of matter that can exist. The implication of these two assertions is that the current scientific activities at CERN, and other labs, to be the first to create synthetic black-hole material are dangerous and could lead to the end of our planet. Of course, even if a black-hole were to destroy the Earth that would not be the end of the Universe. The cycle of Big Bang to Big Bang would be expected to occur into infinity. However, loss of the Earth would be a truly sad event. Fortunately, it is an event that can still be prevented.

## Appendix 1: Scientific American Community

This section is comprised of excerpts from the Scientific American community blog that was started in order to tell the world about this book. Readers of this book were once encouraged to go to the blog where debate had raged over this new model at:

http://science-community.sciam.com/forum.jspa?forumID=300005039

Unfortunately, if one goes to this URL today there is nothing but a bland generic promotion of Scientific American. No mention or credit is given to what was once there. No mention or reasoning is given for their sudden decision to obliterate all of that blog, and what was once a community of bloggers. As of July 1, 2008 SciAm made the "corporate" decision to pull all blogs—including this one. Whatever their motivation, it cannot change what actually transpired over the six months the blog was up, debate ran, and the model stood impervious to the attempted attacks of detractors.

The blog began on Dec 14th, with an introductory article, which was quickly pounced on by particle physicists. The rhetorical war escalated. On Dec 20th, the SciAm blogs were apparently hacked into to stop discussion. When SciAm was back online Dec 26th, I shot back, and posted links for free downloads of the abridged version of this work. On Jan 3rd, hackers vandalized Hasanuddin.org and destroyed link to the free version. My adolescent webmasters put up another link almost immediately. On Jan 4th, the second link was destroyed. Since that day, the free download was made available through a professional web-resource, SendSpace, which had the firewalls needed to deter further hacking. I will not give up. Someone either has to prove my model wrong (which hasn't happened yet despite the amount of time that it's already been freely available), or the actual doomsday sequence is initiated. No matter what obstacles are put in my way, I will adjust and I will fight back.

The loss of the base at Scientific American was a blow. However, it was only a tactical and political loss. The scientific foundation of the model were not affected by this move. If this action was a result of CERN forcing the hand of SciAm, it only serves to show the level of despiration of the theoretical physics community against a new model that seeks to invalidate all of their beloved, yet unverified, relativistic theories and formulae. While the blog was up and running many esteemed physicists did debate to unseat the Dominium model. However, istead of locating flaws, these wanna-be detractors ultimately supplying several new lines of supporting empirical evidence in favor of the new model. The articles in this Appendix 1 were written in direct response to these "detractors" and highlight the new lines of support. The most notable detractor, a scientist from the Tevatron facility ultimately supplied the most concrete evidence. Therefore, this conversation is shown in its own Appendix 2. Neither Appendix has been edited in any way to preserve the integrity of what was actually said.

**Supernova 1987—Proof of opposites repel matter/antimatter interaction**

Unedited segment of a Scientific American blog posted on Dec 19th, 2007

Before I go on, let me sincerely thank "Dan M" for contributing evidence in support of the Dominium hypothesis. He brought to light that there was an event, Supernova 1987, that showed a differential arrival time between neutrinos and antineutrinos—12 seconds. Although this may not seem like a lot of time in someone's life, it is significant. Why would there be a time delay at all? Although this point wasn't presented as support the new Dominium model, it does. In "Dan M"'s original statement, he was convinced that if matter and antimatter truly repel, then the lag-time between the two peaks should have been closer to ten months, though it was never explained how this number was obtained. Therefore, he reasoned, since ten months is "more significant" than twelve seconds, the hypothesis that matter and antimatter repel is incorrect. Apologies to "Dan M," but because the method of obtaining the value of ten months was not defined, it cannot be considered solid. A flaw in calculation could easily explain for the discrepancy between the expected vs. actual gap between peaks, without discrediting the solution that matter/antimatter repulsion was responsible. For example, if the number came from an analysis that supposed that the supernova of origin and the Helios system were the only points in the Universe then the predictions could become skewed and the predictions inflated. Consider the statement that is commonly made, "there is no gravity in deep space." This oversimplification is false. In deep space there will be gravity vectors acting on an object from all other objects within the Universe. However, since there would be other objects within the Universe existing in all directions, the net gravity vector might approach zero. This would be true of all objects. However, because the net gravity vector might approach zero, doesn't mean that it actually is zero or that gravitational interplay is not in effect. Therefore, neglecting the existence of the rest of the Universe alters the situation for calculation purposes away from the actual dynamic of the true natural system being described. The fact that the arrival times of neutrinos and antineutrinos are different, is extremely significant. The fact that the actual gap does not match someone's calculation is not necessarily significant. To explain the gap in the arrival time of the neutrinos and antineutrinos, we need a working hypothesis that is compatible with this observation. The Dominium hypothesis does offer such a response: differential force experienced, the matter-particle would feel a gravitational attraction to the Earth/Helios while the antiparticle would feel a gravitational repulsion, therefore arrival times would differ. Old models offer no such explanation. This is quite strong evidence because none of the old models are able to explain this documented phenomenon.

**All things being equal...but they're not—(Physics vs. Biology)**
Written and posted as a Scientific American blog Dec 31st, 2007—unedited.

A difference of approach: For the stake of creating calculations, there is a phrase that physicists and statisticians use to draw conclusions that is not used by the biologist: "All things being equal." As a person who was initially indoctrinated into the biological sciences, this phrase always seemed a bit funny. How can one claim that all things are equal when it's easy to see that they are not? The other alien contradiction about this phrase is that it is used to abolish nasty interference from anomalous outliers. Rather that consider the significance of an anomaly, with this phrase, they are ignored. This approach, although it conveniently produces mathematical models, is the polar opposite approach to that used by the classic biologist.

With the phrase, "All things being equal," one can easily discount the very existence of all pesky anomalous data to create a mathematical model—so it seems to suit the purposes of the physicist, who prefers to sum up nature in terms of equations and formulae. The equations generated may appear to be impressive. Colleagues might even put a seal of approval on such constructs because of mathematical "flawlessness." But is it flawless? The answer is categorically, "No." Proof of the problems with the mathematical modeling approach is that to describe one particular phenomenon, there might be five or six different math-models that are all apparently without flaw. That can not be true: they can't all be correct; and, therefore, perhaps none of them are correct. Usually, the differences between these models arise from different starting premises and permutations of some line of "all things being equal."

To the biologist all things are never considered equal. The systems studied by the biologist are extremely complex (e.g., endocrine pathways, ecosystems, physiological interactions to a stress/drug, etc.) to attempt to truncate the system [all things being equal] one would warp the entire system into something no longer being what it actually is. Resulting conclusions would be flawed, mistakes made, the patient dies, etc. Documentation of misinterpreting a biological system based on an oversimplification can be found throughout the literature. Consequently, discovery, within Biology, very often comes when somebody realizes what factor is being left out of the consideration, seeing the missed importance, and then telling the world. Therefore, for the biologist, multi-layered complex systems are best approached using a holistic overview, discounting nothing. Because the Universe is an extremely complex system, it follows that the methods of the biologist would be superior to oversimplifications induced by, "All things being equal," and inherently flawed numeric explanations.

In Biology, sameness is of little significance. Rather, it is the outlier and the anomalous that are of supreme concern. The study of mutants has led to a plethora of discoveries in the biological sciences. Anomalous behavior is a sign of possible new disease, discovery, or market capitalization. Biologists stream through mounds of data searching for that one significant anomaly that can lead to something new. Whole orchards are planted in the hopes of discovering one anomalously desired specimen. All peaches are not equal. Genetic engineering and animal husbandry, at their heart, are about identifying differences that can be replicated in future generations for the greater social and market good. Although I have come to love all of the sciences, my undergrad was most richly concentrated in the biological and agricultural sciences. From the very beginning, my training focused on noticing the subtle differences between things. Taxonomy and entomology, for example, are based on seeing the subtle differences between organisms. All bugs are not the same; to insinuate otherwise is foolish. Not even all beetles are the same. Even within one particular species of beetle there may be subtle variations that lead to significant conclusions or discovery. The phrase, "All things beings equal," is of no use in these disciplines.

The first time I heard it said, "all things being equal," I was in my first introducttory physics class. I actually thought the professor was setting the stage for a joke. When no punch-line was forthcoming, I was a bit confused. Such a stage-setting statement seemed like a contradiction from everything else that I had observed or been taught about the natural and biological world, why would the physical world be any different? I raised my hand and asked the professor how one accounts for anomalies. "There are no anomalies in Physics," the professor said with certainty, "Variation in the data is the result of human error." I never liked this answer, but I held my tongue...I was too inexperienced to formulate follow-up questions.

When I came into contact with particle physics and astrophysics, I was still a biologist at heart, even though I had completed many engineering-type and statisticcally based courses in the interim. I was delighted to find (just as I had originally suspected) physics did have anomalies and areas that were not easy to explained under current theories. This was when physics became truly fascinating. While at CERN, instead of joining my friends at the cafe, I opted to go to the lecture hall and did a lot of reading. I wanted to know as much as possible about the current edges of understanding. When faced with an anomaly—that some tried to explain away using an "all things being equal" preface—I only half paid-attention to these lines of thought, because they usually didn't mesh well with more accepted bodies of knowledge. Rather than focusing on the conformist "solutions" (where all things were considered equal), I focused on the anomalies themselves: what were they, what was their dynamic, what evidence lead to their discovery, what might they be connected to, etc.

To me, a biologist/naturalist, the anomalies are much more important than the mush of sameness within any given data-set. As I was exposed to more and more particle and astrophysics, I compiled a list of anomalies from the different lectures, discussions, and books that I read. Whenever I would come across an anomaly that was not easily explained under current theory, I wrote it down and tucked it away:

1) What drives the solar wind in a unidirectional flow AWAY from a giant star—no definitive explanation.
2) Why didn't the entire galaxy collapse into the forming supermassive black-hole—no definitive explanation.
3) Why is the highest concentration of stars/mass, in a galaxy such as ours, closest to the central supermassive black-hole (instead of having an arrangement similar to what we observe in our own solar system)—no definitive answer.
4) Why do nebulae form in one spot of space over another—no definitive explanation.
5) Why is the Universe expanding (if it is all made of gravitationally attractive mass)—no definitive explanation.
6) Why is there an inflection point in the graph of the acceleration of the Universe—no definitive explanation.
7) Why does there appear to be a uniform distribution of mass in the Universe—no definitive explanation.
8) What accounts for the observation of "dark matter" existing between galaxy clusters where no model predicted it to be—no definitive explanation.
9) Why does there appear to be a flat event horizon—no definitive explanation.
10) What caused the Tunguska event, yet left no trace of evidence other than fallen trees—no definitive explanation.
11) If the sun is a positron generator, why haven't positrons been detected in the solar wind by monitoring probes—no definitive explanation.
12) If matter and antimatter were created in equal proportions during the Big Bang, where is the antimatter—no definitive explanation.
13) If antimatter still exists in the Universe, why don't we observe evidence of annihilation events—no definitive explanation.
14) Why was there a documented gap between the peaks of the arrival of neutrinos versus antineutrinos in the Supernova 1987 event—no definitive explanation.

My approach is the polar opposite of the classical physicist. Instead of looking for an equation that aligns with 80%, 90% or even 98% of the data, I sought possible qualitative explanations for the outliers/anomalies. Though some may castigate all qualitative analysis, the fact is: if done correctly, the formal deductive process is very tight and defendable.

Fact: All of the above-listed anomalies are accounted for within the new Dominium model. This alone shows the strength of this new model over the older assumptions and theories. True, there are existing theories, and in some cases, mathematical models to account for each of the above listed anomalies separately.

However, many such "solutions" utilize only-in-this-case tricks to be able to work. Also, these "solutions" are disjointed and unconnected to any other principle or aspect of scientific understanding. For example, the "void of space" theory seems to provide a possible answer for the solar wind driver, yet this "answer" does not fit neatly in with anything else, nor does it explain exactly how the feat it accomplished. To measure the worth of a particular "solution" theory to explain an anomaly I would always judge it against the same set of criteria:

1) How well does this "solution" fit in with more established scientific principles?
2) What proof, if any, is backing up this "solution"?
3) How was the "proof" obtained, e.g., test design?
4) How conclusive is this "solution" (margins and sources of error)?
5) Did the lab obtaining the data set-out to find this result [a bad thing], or was it an unintended surprise [a good thing]?
6) Have the results been independently reproduced with a high degree of certainty?

Using these criteria, the compiled list of anomalous phenomena grew to the 14 listed.

Unlike the previous hodge-podge of disconnected, inconclusive, and only-in-this-case "solutions," the Dominium model provides a smooth, seamless, inclusive, and all-encompassing set of conclusions, which not only explain these phenomena, but predict their existence as they appear.

In the third article of the Scientific American blog series (but appears later in this section), Supernova 1987..., I postulated a "guess" of how the theoretical gap of 10 months was achieved by the mathematical modelers. Actually, it wasn't a real guess, I once had a particlist show me how this number was obtained. The process begins with, "All things being equal, there is no gravity in deep space..." From that point forward, the analysis becomes one of considering only two points: Earth/Helios and the supernova of origin, and their effects on the moving particles. Although doing such a calculation is "easy" when you narrow down the field to only two points, it is neither conclusive nor descriptive of the actual system being described. The fact is, there is a whole rest of the Universe out there. As such, each point of mass in the Universe would inflict its own gravitational force vector on the traveling neutrinos/antineutrinos. Because of this high degree of "gravitational-noise" the effects imposed by the Earth/Helios system would only be expected to play a major part in affecting the movement of the traveling particles when they become very close, conversely the effects of the original system that went supernova would diminish more quickly than "expected." Therefore, actual peak separation would be much shorter than the calculated (and oversimplified) prediction. Depending on the significance of the gravitational-noise a difference of six orders of magnitude (as is the case) is not unreasonable. The only thing that the difference between the actual and theoretical numbers implies is that there was a high degree of error caused by the oversimplification induced by the original, "All things being equal," assumption. The reality is: a 12 second gap is a very long time and it is significant, even though it might rhetorically sound trivial and dismissible. Try it right now: stop reading and look at the second hand of a clock and wait to see how long 12 seconds actually is. Another way to think of the significance of this data is to consider how far apart the different fronts were by the time they reached the Earth. If they were both moving at roughly the speed of light, a twelve second gap would put them 3,600,000,000 meters apart {approx ten times further than the Moon is from the Earth!} That's a huge gap for two different particles that were expected (under more firmly established theories/principles) to be moving at the same exact speed.

Another example of the oversimplification induced by "all things being equal" is the assessment that the Universe is completely comprised of matter. This statement is categorically untrue given what we know of the dynamics of our Sun. We know it is a positron generator; that is an undisputed fact. Therefore, to say that "the entire Universe is completely comprised of matter, and hence, the importance and role of antimatter can be ignored," is a gross oversimplification that is in conflict with known facts. The presence of the positrons of the Sun is brushed aside by the statement, "Well, you need matter in order to have fusion, therefore the matter had to be there first; it's a chicken-and-the-egg thing." Again, such a statement is based on assumption that ignores the possibility that antimatter might be able to undergo fusion (thereby generating electrons) in other quadrants of the Universe. Ignoring the existence of antimatter might appear to make the creation of equations easier, however, by excluding the importance of antimatter huge distortions result.

Mathematical calculations are not the end-all and be-all of scientific inquiry. Ideally, a qualitative structure is first needed that is able to give a narrative of the overall nature before mathematical supports can be constructed. Such was the case with Copernicus who provided a qualitative structure; later, Kepler et al., added the correct mathematical structure. By ignoring the importance of a qualitative structure, mathematical models can be generated that may appear quite sound (Ptolemy) and utilize only-in-this-case tricks (Ptolemy, also), yet describe nothing as it actually occurs, because the premises used are either overly-simplified or wrong.

Contrary to what the particlist establishment might want to believe, even their mathematical models began as qualitative roadmap. In 1967 Andrei Sakharov provided the qualitative map that is been used today. Essentially this conceptual solution is: because of differential decay patterns, the antimatter that must have been produced in the Big Bang "disappeared" leaving us with an entirely matter-only Universe. Once the qualitative road map was established, the mathematical models followed. It is possible to create a mathematical model to justify just about anything, though the more incorrect the qualitative roadmap the more complicated the mathematical model must be. Again, this follows very closely to Ptolemy. For Ptolemy, the qualitative understanding was that the Universe must revolve around the Earth because we are God's chosen. Mathematical models were created that seemed to tie-in "most" of the data. Anomalies, however, plagued Ptolemy (the planets) and only-in-this-case solutions and equations were devised to deal with each planet separately. The parallels between Sakharov and Ptolemy are quite tight.

Now there is a new model—the Dominium. This model is totally incompatible with the established roadmap of Sakharov. However, it is completely compatible with known data and observation. Because it is 180° away from the established direction of Sakharov, it is predicted to be difficult for the establishment to accept—even if it is 100% correct. If this is the correct qualitative roadmap, mathematical supports will eventually be constructed. Time will tell; but if the most ominous implication of this new model is correct, we don't have much time to dally. Debate needs to occur. And, if this new model is actually correct, minds need to change.

The Dominium model is not the result of armchair reasoning, as surely some will claim. The entire model is constructed through formal Aristotelian syllogisms. Although archaic, this methodology is among the tightest and most defendable forms of formal logic. There are only five premises used. From those five springs a construct that produces 47 separate conclusions. Among those 47 are predictions and explanations for all of the anomalies listed previously. Also, supersymmetry between gravitational and electrical systems is catalogued in 10 different and distinct ways.

No, this is not an alarmist paper on doomsday, as others will try to dismiss. However, one of the implications/conclusions of this model is that mini black-holes will be stable. The significance of this projection should be apparent to all: LHC aims to create the first synthetic mini black-hole. Alarming—yes; alarmist—no. Never have I stated that the End-of-the-Earth must happen. Rather, it has been stated over and over that this theory proves that the current direction of the LHC project is based on false assumptions and therefore needs to be reevaluated. The consequences of not considering the implications of this model are truly grave, especially if this new theory proves correct.

No also, to assertions that this model is simply pulp-fiction and an attempt by the author to make money. Although it is correct that this work is being commercially sold/distributed, it was done as a matter of necessity and haste—not for monetary gains or lack of merit. If you analyze the price structure of this book, it was not designed to profit-maximize. Rather, the book was intentionally priced low so that distribution could be what is maximized. The most ominous implication of this book indicates that we (as a planet) may only have five months to change course. If a stable black-hole is produced, then there will be no turning back and an apocalyptic sequence will ensue. Because the normal scientific process of literary review can often take years to complete, the free market was the only option recognized that might succeed in saving us all. True also, that the author had this work categorized by the ISBN agency as both "science" and "science-fiction."

This contradictory classification was done for tactical/pragmatic reasons regarding distribution: there are more sci-fi readers than for hard science. To stop this project, as many people as possible need to read, understand, and act. Also, this classification was done to stay humble. All hypotheses are inherently fictional until they are categorically verified (including the fundamental "CPT Violation" hypothesis of my "opponents.") To call any hypothesis that has not been conclusively verified anything other than fiction is a high act of unscientific arrogance.

Although "all things being equal" has been shown to lead to oversimplifications and false conclusions, this statement can accurately be applied to this internet forum (so long as no more "system errors" occur). "There is an exception to every rule," it is often said. Up until now, I've made a case that "All things being equal" doesn't apply to consideration of extremely complex systems like the Universe and shouldn't be overly relied upon. There is a major exception: deep-down, all people really are equals. Social class, degrees/plaques, income-level, race, and nation of residence are all artificial distinctions. Within our hearts, we are all just as worthy as any other—despite societally imposed restrictions or egotistical inflation. In "true" scientific discussion, this is apparent. Whether the conversation involves a decorated particlist, an undergrad at UC Berkeley, or an unknown troll, like myself, the world community can equally view our words, arguments, and theories as they are posted here on the web. The world ultimately decides what "science" is correct. Therefore, if you can add an insight, article of support, or line of scientific criticism, then do so. Go to the Scientific American community blog-site, find Hasanuddin's blog, and join the discussion. Neither underestimate nor overestimate yourself—but do try to contribute to the discussion if you can. Time may be limited.

A new theory describing the nature of the Universe is of concern to everyone. A new theory that embraces and accounts for all known anomalies is remarkable in and of itself, and therefore increases the need for its attention. When such a theory overturns the apple-cart of prior models, its importance to be discussed increases even further. When such a theory possesses far reaching implications of mortal significance to all the inhabitants of the planet, the importance for actual discussion grows exponentially. Failure to discuss such a new theory is a huge disservice to science, humanity, and one's own integrity—regardless of how distasteful the possible outcomes of such discussion might be.

**Dinosaur problems for Einstein and Sakharov**
Part of a blog posted to Scientific American on Jan 4th, 2008—unedited.

A main problem that exists within the current theories is both large and central; it has to do with the understanding of a single variable. Yet, this variable is not usually center-stage in most people's minds. Because this variable is not on the tops of most peoples' minds, it is easily overlooked or dismissed. As such, when trying to create mathematical explanations, and when "all things considered equal" is pronounced, this particular variant could easily be left out or ill-defined: Entropy.

Webster's defines "entropy" to be: 1) the degradation of the matter and energy of the universe to an ultimate state of inert uniformity; 2) a process of degradation or running down or a trend to disorder. Both of these definitions are challenged by the new Dominium model. In the case of definition #1, the new model provides an alternate prediction for the future course of the Universe. Definition #2 is actually an untrue statement, the new model shows. The fact is, not all systems head toward disorder and this can be easily proven. In science, a particular assertion can be found categorically untrue if a single contrary example can be found within nature. That will be shown to be the case for Webster's 2nd definition for entropy.

Current understanding is that all systems head to a state of increasing disorder. But that is not true. Take for example a tub half full with oil and half full with water. You can insert energy and mix the fluids vigorously. You have just created a "heterogeneous system" of oil and water. If you stop adding energy and isolate the system, what happens? Does it become more disorderly? No. The old understanding might state, "because it is now extremely heterogeneous it is at a state near maximum disorder." Therefore, one might conclude that the system will stay static at this "optimal" high level of disorder, but that doesn't happen either. Is this highly heterogeneous mixture an enclosed and isolated system—yes. Being at or near the state of maximum disorder, according to current understanding of entropy, the system should stay in that form. As we all know, the reverse happens. Instantly, micelles form, they combine, and they self-organize themselves. From what was an extremely heterogeneous system to begin with, an extremely organized arrangement of two homologous layers is formed.

Although there may be some who will not want to accept that such a perfectly mundane system could provide evidence against their "modern" understandings of one of the frontiers of scientific understanding. The fact remains: this is an example of an enclosed and isolated system acting in reverse as to what is prescribed by entropy. Also, because this example is mundane, simple, and easily reproducible it is, therefore, extremely strong refuting evidence against the old understanding of entropy. If this experiment is conducted in space (to rid the gravitational influence of the Earth and therefore densities of the liquids), a self-organization also occurs though the end result is usually a matrix of alternating micelles rather than layers. Either way, the system moved from chaotic to orderly and defies the established understanding of entropy.

Actually, this process is an important and highly studied phenomenon (by people in fields of science other than particle/astrophysics), which can also be considered at the frontier of scientific understanding. In the world of the nanotechnologist, this process is known as "self-assembly." It is a cornerstone to nano-manipulation of surface structures. On the nano-scale, the outcome of self-assembly need not be two separated layers. In fact there are several identified outcomes of "self-assembly" (tubules, micelles, and layers) depending on the conditions created in the lab. The geometry of the subsequent matrix can be predetermined and used in the decision-making of the execution of nano-structure construction. The process of self-assembly is so predictable that there are hopes that it can successfully be deployed in profitable industrial production regimes for future mass-produced nano-engineered products. Documentation of self-assembly is found throughout the modern literature. It is an undeniably important nanotechnology tool whose applications are still being discovered.

To the nanotechnologist, concerns related to entropy do not enter in to their "big picture" and how self-assembly fits into their work. Rather the nanotechnologist is focused on using the tool of self-assembly in order to achieve a desired practical and engineered outcome. Therefore, the focus of research is to understand the dynamics of this new and powerful tool. Because there is limited dialogue (accomplishments, processes used, and possible new technology/discovery) between disciplines, the tools of the nanotechnologist would not be often discussed with someone more concerned with the workings of entropy. This explains why a set of knowledge could remain "hidden" in one subgroup of science, yet not be noticed or utilized by another.

Even the name, "self-assembly," flies defiantly in the face of common understanding of entropy. One understanding of entropy is that a system can only become more orderly with input of energy from the outside. Yet, as the name implies, for self-assembly, the energy to organize is drawn from within the system's "self," not from the outside. Also, the word "assembly" is the antithesis of the "degrading" aspects of the old understandings of entropy as defined by Webster. Despite these contradictory terms, this name was aptly chosen by the nanotechnologist; it is a practical description of a tool where a given system will self-organize in a predetermined manner.

Another example (or form) of systems that self-assemble is crystallization. In this process, exacting order is given to particles that were once organized randomly. Again, depending on what type of crystallization you examine, outcomes are predictable and precise, especially on the atomic level. A disordered system becomes more orderly. Crystals, again, are nothing new and are documented throughout the literature. However, one might discard this example because it represents a phase-change. Though in true self-assembly of the nanotechnologist, there is no phase-change.

Not so surprisingly, a major difference, between the new Dominium model and the old models, is self-assembly. Before that can be shown, a quick explanation of self-assembly needs to be given. For true self-assembly to take place, there needs to be two different types of particle. The two types should have properties (for nano-structures and crystals: electrical) that are distinct from one another. Because of the different properties there will be interaction. The interaction consists of pulls and pushes between neighbors (both adjacent and far away). As a result of proximal asymmetries between the pulls and pushes, the system is forced to re-adjust. This readjusting process is what's called "self-assembly." And the resulting system is more orderly and stable than it had been prior to the readjustment.

In commonly observed self-assembly, the distinction between particles that had led to the readjustment was electrically driven. This fact does not preclude the possibility that self-assembly could occur to systems that are made of two distinct particles other than electrically based ones, i.e., gravitational.

The Dominium model makes the assertion that self-assembly occurred at the galactic level immediately after the Big Bang and is driven by competing gravitational forces. The initial result of the Big Bang was to create an extremely heterogeneous system comprised of equal amounts of matter and antimatter—two distinct particles—extremely analogous to the highly mixed container of oil and water. The central hypothesis of the Dominium model is that matter and antimatter repel one-another; also, it is a given that like-attracts-like in 100% accepted (all known matter-objects are attracted to other matter-objects given Newton's Universal Law of Gravitation). With the central hypothesis that gravitational opposites (matter and antimatter) repel, we can now envision the Universe immediately after the Big Bang to be a system very close to those created by the nano-technologist. At the galactic level, gravitational pushes and pulls would be felt between neighbors, near and far. Because of asymmetries, the system would readjust. The result of the readjustment would be orderly, and possibly predictable.

Pleasantly, that is what we observe: mass distribution of the Universe conducted by SDSS and other labs have shown that the distribution of mass is extremely uniform. In fact, it is much more uniform than many experts had calculated that it "should" be. A semi-rigid crystalline structure would also be expected to stretch the event horizon flat in all directions. Again, this is exactly what has been observed.

An irony of the Dominium solution is that a prerequisite of self-assembly is an asymmetry between/among the forces felt by the particles of the system. The asymmetry is what drives the system to readjust. The outcome is a more symmetric and orderly matrix. This is ironic because the central hypothesis of the old theory (CPT Violation) is one where absolute "importance" is put on asymmetry. Ironic, because the Dominium concurs: asymmetry was crucial, but only to set the stage for symmetry.

The title of this discussion refers to the problems resulting from flawed entropy understanding as dinosaurs. Indeed they are. Like brontosauruses, these problems are huge and are embedded in the fundamental underpinnings of the old theories. Like the dinosaurs, these problems predate us. And also like the terrible-lizards, these problems should become extinct. All the old models that have relied on the old flawed understanding of entropy should also be fossilized and buried; unless of course, those older understandings are also compatible with the new understanding of entropy that incorporates the documented process of self-assembly.[1]

[1] A person hopped onto this article and commented that I was wasting my time trying to prove something that was already "well-known." He stated that up until the end of the 20th century it was held that entropy always naturally sought a more disordered state for all closed systems. However, in the 1980's it was shown conclusively that is not always the case. He then referred me to a link of a website that shows just that: www.entropylaw.com For your contribution, thank you "Chuck Darwin."

**Reverse-gravity: An only-in-this-case trick supporting the Dominium model**
Unedited blog article posted on Scientific American January 2008

As mentioned in a previous discussion, "all things being equal," is a phrase that hinders the development of Physics past its current position of understanding. Another case, not yet discussed, where this phrase has hampered understanding is with the approach to the events immediately after the Big Bang. There is a well-known acceleration curve that shows what is known about the dynamics of the acceleration of Hubble expansion of the Universe as it has been observed. This graph is usually cut up into three distinct subsections: initial expansion, inflection point, and current expansion curve. Rather than view this curve in its totality, mathematical modelers instead chose to explain each section separately. Many explanations for the initial expansion rely on an only-in-this-case trick of "reverse-gravity." Because researchers often work independently of one another, subsequent portions of the expansion curve often ignore reverse-gravity, "all things being equal."

Although I do not know who to give credit for the introduction of the necessity of reverse-gravity to explain the initial phase of the accepted acceleration curve of the expansion of the Universe, reference can be found throughout the literature, e.g., Guth. This only-in-this-case trick is widely accepted to be necessary to explain the initial portion of the graph (rapid expansion). However, when you view the explanations for the point of inflection and the current phase, mention of reverse-gravity is lost.

The Dominium model asserts that the mathematical modelers who originally employed reverse-gravity are the ones who were correct, not the subsequent ones who ignored its importance in their models. The Dominium model asserts: reverse-gravity would be the opposites-repel interaction between matter and anti-matter. The Dominium model asserts that this interaction combined with like-attracts-like interaction lead to self-assembly and movement of the Universe to a more organized state (see Dinosaur problems…). Once self-assembly complete the most dynamic effects of the presence of reverse-gravity would be hugely diminished, yet it would still be present.

Let me pause. Mathematical modeling is a powerful tool for analyzing a sub-section of a system. Market research, agriculture, and industry have proven again and again that mathematical modeling, if done correctly, has a profound impact on the bottom-line. If you know enough about the system, a strong mathematical model can be produced that can predict how the system will react to future stimuli. But you have to know "enough" about the system in order to make the model correctly. In the case of the three subsections of the Hubble expansion acceleration curve, the modelers who analyzed the initial rapid expansion knew "enough" about the system to conclude that there must have been "reverse-gravity" acting. For those working on fitting models to the inflection point, they don't know "enough" about the system and their mathematical models are not accurate descriptions of the dynamics of that period. Because of the diminished role of reverse-gravity (matter/antimatter repulsion) after the completion of self-assembly, it is possible that the mathematical modelers could have produced models that come close to describing the actual system, yet ignore the presence of reverse-gravity, "all things being equal."

The inflection point and modern portion of the expansion curve are also explained by the Dominium model. According to the new model, the inflection point involves other occurrences than just simple self-assembly. These other processes are so profound that the whole phenomenon together is given its own delineation: the Dark Event. However, the concert of interaction is quite complicated and would detract from the present article if explained further. A full account of the Dark Event is contained inside of both available versions of the Dominium (conclusions 13 through 20).

In the case of reverse-gravity, this concept is probably ignored because the accepted practice of a disjointed approach/overview between labs and researchers. For example, the MIT PhD who triggered this book described his line of research as "only concerned with the inflection portion" of the of the expansion curve. I hope he was being overly simplistic in his description. However, many labs do work in isolation of the discoveries being made by others, for example, how the old understanding of entropy has been allowed to stand for so long despite decades worth of contrary evidence produced by the nanotechnologist (see previous discussion). By not considering the integrity of the entire curve holistically, a factor that was essential for one part of the curve can be left out of consideration for later parts. True a mathematical model can be generated that creates a curve that aligns with the known curve, however, that does not imply that the mathematical "solution" is correct or conjoined to the other parts of the graph.

Because the degree of curvature of space-time would be highest when dominia are closest to one another, acceleration would have been at its maximum earliest on. This accounts for initial aspects of highest outward acceleration of the embryonic galaxies, as is shown in accepted representations of Hubble expansion. During the earliest moments, repulsive gravity between matter and antimatter structures would have been in its most ballooned state because of proximity. As distances increased, the rate of acceleration due to gravitational repulsion would have decreased. [Let it be noted that annihilation events that would have been expected during this period that could have contributed an effect on the acceleration rates of the embryonic matrix.] The sudden reduction of the acceleration rate noted on accepted graphs of Hubble expansion can initially be explained by a combination of two factors: the proportionality of F changing from $d^{-7}$ (or more extreme) to $d^{-2}$ and the advent of immiscibility.

As evidenced by multiple papers, in order to obtain a mathematical model that can produce the type of initial rapid expansion of the Universe, some form of reverse-gravity must be inserted. The insertion of this concept aligns completely with the Dominium hypothesis of matter/antimatter repulsion. However, there is no explanation under the old established theories that can account for any form of reverse-gravity. Therefore, the mathematical models, which have been produced to describe the rapid expansion phase of the Universe immediately after the Big Bang, provide strong evidence in support of the new Dominium model.

---

## Of Copernicus, Ptolemy, and the solar system

Unedited version of a blog article that was posted January 16, 2008

To understand the dynamic of the current impasse between the Dominium model and the particlist establishment one can look to history to find a very similar conflict. A strikingly similar scenario was created when Copernicus appeared with a revolutionary new qualitative understanding of the solar system. The reactions of the scientific community of that time strongly mirror the reactions seen today against the Dominium. In fact, the very nature of the arguments of current detract-tors run in parallel to those used against the Copernican heliocentric model.

Before the arrival of Copernicus, the scientific community had reached consensus that the Universe rotated around the Earth. Strongest evidence of the old model came by observation of the Sun and Moon, which "obviously" circled the Earth. Evidence was also sighted in the way that the stars of the Universe seem to predictably rotate around the Earth, as the night unfolds. There were, however, anomalies. Venus, Jupiter, Mars, Saturn, and even Mercury were well known to the ancients. However, their observed motion did not fit easily into the model of rotation around the Earth. The word "planet" even comes from the ancient Greek word "wanderer." To account for the anomalous wandering of the planets, Ptolemy and his followers drew up complex equations. The type of mathematics used by Ptolemy was extremely complex and multi-layered. Each planet required its own only-in-this-case tricks in order to achieve "solutions" that appeared to match the observed behavior. The type of math used by Ptolemy was distinct and not applicable for any other purpose. Even though only a few individuals could actually claim to understand the explanations of Ptolemy, the scientific community reached a tight consensus. Ptolemy's model was further enforced and used by the church to confirm the belief that we were God's chosen and therefore at the Center of all Creation. After all, despite the complex explanations for the planets, who could deny the behavior of the Sun and Moon?

The complex math displayed on the chalkboard office-displays and the computer simulations of many modern physicists around the world highlight parallels between this historic example and the current situation. As mentioned, in the days of Ptolemy, the Earth was considered the center of the Universe and the behavior of the planets (wanderers) presented anomaly to this theory; which parallel the anomalies of today's theories (supermassive black-holes located in the most densely populated area of the galaxy yet hardly feeding, a solar wind that defies easy explanation as a unidirectional flow away from the Sun, the absence of antimatter annihilation events despite accepted notions of pair-production, etc). To account for the anomalous wanderings of the planets extremely complex equations were devised using only-in-this-case tricks; such are the "solutions" for today's anomalies. The prowess of the scientist of Ptolemy's day was often based on the complexity of the equations that were developed; similarly, modern theorists boast of solutions using multiple parallel Universes, multiple ill-defined dimensions, and theoretic types of force and energy (e.g. "dark energy") that apply only to the models being discussed.

The math employed by Ptolemy (and modern particlists) does not crossover or have any apparent application for other branches of science. In both cases, this fact was/is brushed aside. For Ptolemy, it was assumed that because they were dealing with heavenly objects, the math used to describe their behavior was necessarily complex. Similar rationalizations are employed today: modern particle physics, it is assumed, is elite from the other sciences because of the high energies and minute scales being examined. Of course there is no crossover of relativistic applications to common low-energy macro-sciences, such as Biology, Chemistry, or basic Physics.

The acolytes of Ptolemy claimed victory for solving the questions of the Universe because their equations seemed to provide answers that matched the observable behavior of the planets. Similarly, today's particlists claim to have unlocked the answers because their computer simulations appear to match the particular anomalous observation they were hoping to explain.

When Copernicus presented a clean, simple, and qualitative explanation, he was dismissed as a heretic, even though he had been a priest himself. The acolytes of Ptolemy had convinced themselves that they alone had the true understanding of the Universe, and therefore of God. Galileo and others who embraced the Copernican model were derided as blasphemous, tortured, imprisoned, and shunned. The old guard was loath to abandon the complex equations that they had spent so long developing. Rather that consider the new model, they held onto the old even tighter than ever. Instead of discussing the new heliocentric model, they used their own mathematical proofs as evidence of their absolute correctness. It took years before the scientific community accepted the truth and beauty of the new model presented by Copernicus. Essentially, the old guard of Ptolemy supporters needed to die off before the Copernican model could be fully adopted by the scientific community.

The historic similarities are striking. The currently held old theories contain multiple dead-ends and are confounded by anomaly (What drives the solar wind?—Why didn't the galaxy all collapse into the supermassive central black-hole?—What is responsible for the gaps measured between the arrival of neutrino versus antineutrino from Supernova 1987?—What existed before the Big Bang?—Where is the current Universe heading?—What is dark matter?—Why is the Universe evenly distributed?—Why do we see a flat event horizon?—What was the Tunguska event?—Why is the acceleration curve bent?—Why is the Universe expanding?). Wow, that is a lot of unanswered questions for the old theories to have unaddressed! To account for these anomalies, there are many only-in-this-case equations have been devised which employ extremely complex types of math applicable nowhere else in nature (very similar to Ptolemy's "solutions.") Some of the elite in this field even boast about the complexity of the solutions, rather than their simplicity (also similar to Ptolemy). They have devised explanations for one or two of the listed anomalies that employ 6, 10, 12, 22 parallel dimensions to get close to what we observe. However, even these attempts only address a subset of the listed unexplained anomalies.

Now I arrive on to the scene, an outsider, presenting a qualitative model, which is strong because of its simplicity, which does not rely on the use of multiple dimensions, parallel Universes, or bizarre only-in-this-case type of conditions, and which provides a seamless explanation of all observed phenomena from Big Bang to the next Big Bang. Like Copernicus, the model I offer is qualitative and not of the normal variety. Therefore it must be discussed.

A particle physicist once told me that Copernicus wasn't a real scientist, "After all, Copernicus didn't really provide the world with anything special. All he had was a qualitative description...anyone could have said what he said." Although this man's oversimplification of history is kind'a correct, no-one else had voiced the ideas that Copernicus brought forward. Therefore, Copernicus must be considered a truly great man of our common history. If you don't want to call him a "scientist," then call him a "visionary" or a "philosopher." No matter how you phrase it, his contribution forever changed the course of science. I, too, am not a normal "scientist." Regardless of my classification, the model that I am advancing is strong and has yet to be successfully challenged.

Though I was never forced to learn the equations of Ptolemy, my cousin, Lisa, did. She said it was the most confusing, complicated, and convoluted semester she ever experienced as a math major. Similarly, try asking a "modern" particlist try to explain a celestial anomaly, you'll get one of the most convoluted, complex, and twisting explanations, which is filled with only-in-this-case type of applications. If you then ask for a simplification of what was just said, they will just you a blank all-knowing stare and reply, "That was simple." To me, such an answer doesn't cut it. All the sciences I've ever studying can be couched into more simple language—except particle physics. Why should it be different? The all-knowing rebuttal from the particlist is, "Oh, well these are much more complex sorts of matters, of course the answer is going to be complex." Ironically, this justification echoes of the words of acolytes of Ptolemy. As previously stated, it was argued then: since the movement of the planets involved the "Heavens" (i.e., closeness to God), that the equations and explanations were necessarily complex. Any fool could see that...except Copernicus.

This newly published Dominium model is a qualitative thesis by its very design. That is not to presume that, if it is correct, there will not be future quantitative explanations of this same model (similar to how Kepler, et al., refined and quantified the Copernican model). Despite its maverick and qualitative nature, it employs formal Aristotelian syllogisms, which are among the tightest forms of argumentation that has ever been developed.

There is a huge difference between today's conflict and the one that ensued a-round Copernicus—the implication of apocalypse. I don't like this aspect any better than anyone else, yet it exists within the model. However, this model does not present apocalypse as a certainty for the immediate future. Rather, we are at a crossroads: do we go ahead with LHC and potentially create the Beast that will consume the planet, or do we veer off this course, postpone/cancel the experiment, and consider the new Dominium model rationally? Because this is a crossroads, haste is necessary. Apathy and inaction will favor the status quo and LHC will be fired-up on schedule. We don't have time for a long drawn-out impasse, nor for current particlists to die-off before this model can be taken seriously. Unlike the days of Ptolemy, we don't have the luxury of years to wait for real discussion; we only have months. The only prudent course is that LHC start-up must be, for the time being, postponed so that the merits of the two models can be decided. Debate needs to take place.

Another huge difference between today and the days of Copernicus is the internet. Although there are more people living on Earth, we are much more interconnect-ed. The abridged half-version of the model was made available for anyone to retrieve at a click of a mouse. This full version could be purchased by anyone at an online bookstore. And the Scientific American blog was visible to the entire world. In addition, people can contact each other, even at great distances. The abridged version can be emailed to friends and family, who can in turn pass it on. All of the people who have been made aware of the new Dominium model can contact their political representatives, university professors, and Ministries of Science. Today, unlike the days of Ptolemy, the particlist establishment can be bypassed, even if they choose to continue to stonewall. The LHC start-up can potentially be postponed/stopped via political pressure, but it will take the efforts of a lot of "common" people to achieve this goal.

Please if any of the contacted experts respond to you with a personal email rebutting the Dominium model, encourage them to post their rebuttals on that blog. If they do not, push them. If they continue to refuse, paraphrase their rebuttals and post them to Scientific American yourself. If there is an Achilles Heel to the Dominium model, then let it be exposed quickly and to the entire world so that LHC might continue unimpeded. Conversely, if the rebuttals and/or assurances are flawed, then let those flaws be exposed quickly so that LHC can be stopped before it is too late. If an expert refuses to post their rebuttals onto this blog, it can safely be assumed that such comments were merely invalid pseudo-scientific rhetoric.

Also unique to this situation is the fact that there is preexisting opposition to LHC. I have come across two organizations on the web (LifeBoat and LHCdefense) that appear to have as their mission stopping the LHC project. Their reasoning was based on concerns in place even before the advent of the Dominium model. I hope the new model can help them in their efforts. Although I would not recommend overreliance on either of these organizations—there are jobs that we each can and should do toward this goal—however, these organizations might offer useful resources. Whatever you choose to do, follow your heart and remember the children.

This battle between models (Ptolemy vs. Copernicus) is often used to show a dispute that can be resolved using Occam's razor: when confronted with two incompatible models that explain natural phenomena, the simplest answer is most probably correct. If this philosophy holds true, then the Dominium model is highly favored. Although to the reader, some parts of the new model may appear technical, the level of complexity is nowhere near that of the current models dependent on relativistic math, multiple dimensions (more than four), parallel Universes, and/or strange partially defined forms of energy. In addition, the lack of anomaly between the Dominium model and observed natural and experimental data provides the strongest evidence of both its simplicity and its validity.[2]

---

[2] In the comment thread to this blog-article the point was raised that Copernicus was branded a heretic and Galileo mistreated due to religious intolerance because of the birth of Lutheranism rather than scientific intolerance. I would argue that both types of intolerance occurred; a common denominator between people is intolerance toward ideas that conflict with one's own. This is especially true when someone believes that they possess absolute truth and understanding. This is true for both religious and scientific zealots. Therefore, it is predictable that there will be those who reject this body of work because I am a Moslem who embraces the validity of the other religions. Oh well.

**Erasers**

Unedited blog article posted on Scientific American in January 2008

I have been trying to understand nature and crack puzzles for a long time. One humorous anecdotal mystery I've observed is that the offices of physicists "look" much different that those of the biologists with whom I had worked. Though both type of scientists do have the expected four square walls, large messy desk, and shelves of books, Physicists usually have something else—a chalkboard full of equations and scribbles. True, I had seen chalkboards (or whiteboards) in the offices or meeting areas of professors I knew during undergrad at UCDavis—but there was a big consistent difference between the boards of the Biologist versus the boards of the Physicists I've visited at CERN, Northeastern, and elsewhere. The biologists' boards were almost always wiped clean, while the physicists' boards were crammed with alien looking symbols, equations, and diagrams.

I remember very distinctly being surprised the first time I ever walked into a "true" physicists office and saw his blackboard. I was curious. I remember asking what the equations meant. The answer is something I'll never forget, "Oh, it's just a little something I was just working on, but I got stuck." I asked for it to be explained further, because I'm always interested in trying to crack a puzzle, and I thought that I might be of assistance. "Oh, it would take to long to explain," was the chuckled reply, "besides, I don't think you have the background." Again, I'll never forget those words. I don't like being dismissed, nor do I appreciate being called stupid (even in an offhanded manner) especially when I haven't been given a chance.

As time went on, I noticed something else that was distinctly odd about these office chalkboards—they were never erased. The physicist had been "kidding me" when he told me that it was something he was "just working on." In all probability, those scribbles had been on that board for months, if not years. I wouldn't even be surprised if those same scribbles are still on that same board today, seven years later. This observation became apparent on several occasions when I happened to be in a physicist's office and I posed a question that would have been most easily explained through equation or diagram. On one occasion, this same physicist spent a long time searching for a clean piece of paper so that he could make his point (rather than erase part of his "display-board"). Being the cynic that I am, I suggested that he just use the chalkboard. The glare that I got from my little quip almost made me laugh, but I maintained my straight-faced innocence. "Sacrilege," the stare seemed to say, "Erase my chalkboard?"

Here in lies a big difference between the Biologist and the Physicist. The chalk boards at UCDavis are wiped clean so that they are ready to be used again. If someone does forget to erase a diagram, biochemical pathway, or equation, it doesn't stay up for very long because the board will soon be needed for other purposes. The Biologist's chalkboards are dynamic constantly used tool, and therefore kept clean for future investigations. However, the boards in the offices of many physicists are static. They are for display purposes only. Rather than being a working tool in the educational or research endeavor, they are designed to impress visitors and provide the illusion of superiority and intellect. They function in a very similar way to a Picasso hanging in the office of a businessman. Visitors are impressed, and perhaps ready of submit to the authority of this individual because awe induced by the wall-hanging. But, they are superfluous and inconsequential to matters of daily business. Here in also lies the Achilles Heel that has bedeviled physicists for so many years. Unlike a Picasso who's abstract presence helps open up the mind, staring at the same dead-end equations day after day and year after year would be expected to cause the mind to close.

One might say that I am making a flaw in my efforts to win over converts to the new Dominium model by my words of insolence directed against the very people who's field I wish to correct. That may be, however, the words of Kuhn predict that these people will resist this new Dominium solution no matter how it is presented. I firmly believe that the only way that the "top physicists" of the world will be able to accept the truth and beauty of the new Dominium model is if they first wipe clean their chalkboards. Besides, acceptance of the new model need not begin with those professionals who have dedicated many years to the old theories that have all led to dead-ends, unexplainable anomalies, and reliance on bizarre math. Such people have vested interests in the old theories. After all, the scribbles on the blackboards in their offices have helped prop up their sense of self all these years. Rather, acceptance of the new model will probably come from people in more dynamic fields (e.g., physical chemists, nanotechnologists, and/or biologists) who are more used to considering outside options from fields other than their own to solve nature's puzzles.

For example, the phenomenon "self-assembly" is core to the Dominium's explanation to account for how the Universe became arranged into the apparently uniform manner that we now observe. This concept is a relatively new idea that is currently investigated by the physical chemist and nanotechnologist. Probably the most remarkable aspect of self-assembly is the fact that it defies the accepted notions concerning entropy—rather than the system becoming more disordered, it becomes more orderly. Although this fact is widely accepted in some circles, it has yet to penetrate into all fields of study. To accept the validity of the Dominium model, one must first accept the reality of self-assembly. To accept self-assembly, implies that one must first wipe clean pre-existing notions concerning entropy. To question the assumptions concerning entropy, means that the chalkboard diagrams in the offices of theoretical physicists MUST be wiped clean because the old assumptions of entropy are foundation components of these equation-display diagrams. Because awareness of the phenomenon of self-assembly has been around for some time, this chain of deduction should have led to the abandonment of the old models and equations long ago, but it didn't. Rather the discoveries of the physical chemist and nanotechnologist are ignored or deemed inapplicable to the understanding of gravitational systems—somehow elite and apart from the other sciences.

The complex math employed in the chalkboard displays of many physicists around the world highlight a historic example that parallels the current situation. In the days of Ptolemy, the Earth was considered the center of the Universe. The behavior of the planets, or wanderers, presented anomaly to this theory. To account for the anomalous wanderings of the planets, extremely complex equations were devised. The prowess of the scientist of the day was often based on the complexity of the equations that were developed. It was assumed that because they were dealing with heavenly objects, the math used to describe their behavior was necessarily complex. When Copernicus presented a clean, simple, and qualitative explanation, he was derided as a heretic. The old guard was loath to abandon the complex equations that they had spent so long developing. It took years before the scientific community accepted the truth and beauty of the new model, presented by Copernicus.

The historic similarities are striking. The currently held old theories contain multiple dead-ends and are confounded by anomaly (What drives the solar wind?—Why didn't the galaxy all collapse into the supermassive central black-hole?—What is responsible for the gaps measured between the arrival of neutrino versus antineutrino from Supernova 1987?—What existed before the Big Bang?—Where is the current Universe heading?—What is dark matter?—Why is the Universe evenly distributed?—Why do we see a flat event horizon?—What was the Tunguska event?—Why is the acceleration curve bent?—Why is the Universe expanding). Wow, that is a lot of unanswered questions for the old theories to have unanswered! To account for these anomalies, there are many only-in-this-case equations have been devised which employ extremely complex types of math applicable nowhere else in nature (very similar to Ptolemy's "solutions.") Some of the elite in this field even boast about the complexity of the solutions, rather than their simplicity (also similar to Ptolemy). They have devised explanations for one or two of the listed anomalies that employ 6, 10, 12, 22 parallel dimensions to get close to what we observe. However, even these attempts only address a subset of the listed unexplained anomalies.

Now I arrive on to the scene, an outsider, presenting a qualitative model, which is strong because of its simplicity, no use of multiple dimensions or bizarre only-in-this-case type of conditions, and a seamless explanation of all observed phenomena from Big Bang to the next Big Bang. Like Copernicus, the model I offer is qualitative and not of the normal variety. History shows that Copernicus' life was made miserable by those that he tried to correct [as evidenced from my reception on the SciAm blogs and the subsequent hacking] my model is receiving similar "consideration." Although I do not relish the idea that my life could be tortured in a manner similar to the reception given to Copernicus, the model that I have published is clean, seamless, and not confounded by anomaly or dead-ends. Therefore it must be discussed.

Copernicus was locked up, physically tortured, shunned, and exiled by members of the scientific community.[3] He was derided as a heretic, even though he was a priest. Hopefully, society has evolved since that time.

The fact is: we all need erasers, because we all make mistakes and sometimes operate under false assumptions. The amount of chalk dust that accumulates at the bases of my boards is quite remarkable. Many of the scribblings that get placed on my chalkboards start as mistakes that are eventually corrected. But that's the process of learning, isn't it? Whether or not the new Dominium model is right or wrong, it is a new approach that needs to be discussed rationally, scientifically, and respectfully (no more cowardly or rhetorical attempts to hack, squelch, or silence). Only by being open to debate can science advance most effectively.

However, there is a huge difference between today's conflict and the one that ensued around Copernicus: the implication of apocalypse. Trust me, I don't like this aspect any better than anyone else. Yet it exists within the model. Because of this all due haste is necessary. For now, LHC start-up must be, at least, postponed. The debate needs to take place.

---

[3]This is a mistake that I was handily chastised by one of the commentors. I was confusing Copernicus with Galileo who was indeed imprisoned, shunned, etc for his support for and efforts to advance the Copernican model. Copernicus died peacefully long before his model was ever seriously debated. I freely admit that I too make mistakes.

## Tainted and messaged data

Unedited blog article published in January 2008

"Ninety-five percent of all experiments are dead-ends," said my best friend Cathie while I was at UCDavis. She was working on her masters' thesis. The hypothesis she was exploring, "Do alfalfa sprouts induce Lupus in rabbits," was not going well. I had been helping her go through her chores in the sterile lab: feeding the bunnies, drawing blood from their ears, and preparing the samples for quantitative spectral analysis. Though I wasn't at all interested in either bunnies or the outcome of her study, I helped her because I knew that the quicker she could get through these chores the sooner we could go out dancing, eating, or whatever. However, her statement was quite interesting, and I made her explain.

In a very matter-of-fact way she said that most hypotheses turn out to be false; most lines of inquiry turn out to be dead-ends; most labs work for months on a product only to find out that the side-effects are worse than the benefits. Yet, when you look through the literature, you'd never know that—everyone claims to have discovered something of worth. Why? Because if people honestly reported that the experiment was a flop or that the time spent had only verified a dead-end, then the people funding these labs would demand a refund. How can you refund what's already spent? Also, the researcher will soon need to make a request for more funding; you can not ask for funding and expect to receive it if you are perceived as a loser. No, instead you use big words, flip the data this way and that, and spin circles around the people who give you funding. Most of them don't understand that much science, and if you make it technical enough and use enough equations, "they'll glaze over like a donut" and hand you whatever you want. "You see, my study," she said, "is only proving that alfalfa sprouts are good for bunnies—which is nothing. Between now and six months from now, I've got to figure out something that will justify me receiving a master's. It ain't gonna be easy."

I originally thought Cathie's sentiments were a bit off base—I had always been taught and believed in the integrity of science. However, later in my schooling I had a fantastic and cynical Ichthyology professor who presented a lecture that mirrored what Cathie had told me in the bunny-barn. At the end of the lecture, he gave us an extremely odd assignment: comb through the scientific journals and find five articles in the fisheries literature that possess inconclusive results, messaged data, or are otherwise show nothing new. That afternoon I went to Shields Library and began looking for articles related to fish. Finding ones with flaws wasn't hard. I only needed to look through six articles to find five that met the criteria of the assignment. It was a truly eye-opening experience.

All of the papers that I included in my write-up were published in esteemed scientific journals. All of these papers had under-gone "peer review." Yet all appeared to possess blatant weaknesses, and the conclusions that were drawn could not be considered solid. This was very perplexing, so I took it upon myself to ask the professor to explain how such papers could have been published in the first place. He laughed and said, "It's all a game. The people who act as reviewers are in the same boat as the researcher whose paper is being reviewed. They all know each other. Papers are often passed on for publication, even when it is clear that they are inconclusive, because the reviewer knows that someday the tables might be turned." Although this made sense, from an interpersonal standpoint, it seemed hypocritical to the spirit of scientific inquiry.

From that point on, I became very cynical of all new papers. I could never again take the conclusions of research on face-value alone. I retained this cynicism as I advanced and studied the other sciences. It is with this cynical bent that I compiled the list of modern physics anomalies that cannot be explained easily given the theories that attempt to describe them:

a. What drives the solar wind in a unidirectional flow AWAY from a giant star—no definitive explanation.
b. Why didn't the entire galaxy collapse into the forming supermassive black-hole—no definitive explanation.
c. Why is the highest concentration of stars/mass, in a galaxy such as ours, closest to the central supermassive black-hole (instead of having an arrangement similar to what we observe in our own solar system)—no definitive answer.
d. Why do nebulae form in one spot of space over another—no definitive explanation.
e. Why is the Universe expanding (if it is all made of gravitationally attractive mass)—no definitive explanation.
f. Why is there an inflection point in the graph of the acceleration of the Universe—no definitive explanation.
g. Why does there appear to be a uniform distribution of mass in the Universe—no definitive explanation
h. What accounts for the observation of "dark matter" existing between galaxy clusters where no model predicted it to be—no definitive explanation.
i. Why does there appear to be a flat event horizon—no definitive explanation.
j. What caused the Tunguska event, yet left no trace of evidence other than fallen trees—no definitive explanation.
k. If the sun is a positron generator, why haven't positrons been detected in the solar wind by monitoring probes—no definitive explanation.
l. If matter and antimatter were created in equal proportions during the Big Bang, where is the antimatter—no definitive explanation.
m. If antimatter still exists in the Universe, why don't we observe evidence of annihilation events—no definitive explanation.
n. Why was there a documented gap between the peaks of the arrival of neutrinos versus antineutrinos in the Supernova 1987 event—no definitive explanation.

There do exist papers that claim to have found answers to some of these phenomena. Yet, such attempts only account for a subset. Of the ones that possess lab-generated "proof" and accompanying scientific journal articles published, one must maintain skepticism: margins of error are high, set of positive observations are low when compared to overall events, and, worst of all, the labs in question were trying to produce the results that they claimed to have obtained.

Why is the fact that the lab in question was trying to find the results reported the worst aspect of the experiment? To me, it all boils down to sick bunnies. Success is the best type of result that one can achieve in an experiment. When a foregone conclusion is not successfully achieved, the social conclusion is the experiment (and the experimenter) are failures. For Cathie, she had several sets of equally healthy bunnies. Regardless of that fact, she believed that she had to somehow show that the alfalfa sprout fed ones were worst off or she'd jeopardize getting her Masters diploma. For a multimillion-dollar lab, the stakes are even higher.

When someone truly believes that a particular outcome must occur there is also a propensity to see things that aren't really there. For example, take visitors to Loch Ness or Area 51. These people are so convinced before arrival that they are either going to see a lake-serpent or aliens that even the slightest flicker or glare can led them to jump to the conclusion that their trip has been a success—what they believed from the outset is true. Such tourists are not scientists. However, scient-tists are people—not all that different from the tourists. We are all driven by hopes, emotion, and, quite often, ego. Therefore, not only is there the external pressure for a lab to stand behind inconclusive results to justify funding, but there are also strong internal pressures for the individual to see what they wanted to see.

In other cases, researchers have published papers that are overtly falsified. A recent, highly-publicized example of this occurred when a South Korean researcher published a bogus account of being the first to successfully clone another human. This paper rocketed him to stardom and he became a national hero. However, because this field is highly competitive his paper was not taken at face-value. Eventually, it was found that the data was fabricated. Since then he has been disgraced and shunned by his nation. What would motivate him to do such a thing? Perhaps he felt that he wouldn't get caught. Perhaps he had done it before, and did not get caught. Perhaps he felt that it was only a matter of time before someone else claimed the prize that he coveted. Whatever the case, he was blind.

How does this relate to the new Dominium model? The point is: the main compet-ting theory to the Dominium model is that of CPT Violation. Both the Dominium model and CPT Violation can claim one example of "strong" supporting evidence. For the Dominium, it is the observation of the gap in the arrival of neutrinos vs. antineutrinos from Supernova 1987; for the CPT Violation hypothesis it is the apparent asymmetric decay of kaons reported from the BaBar facility. As evidence, the arrival gap noted in Supernova 1987 is far superior to the apparent asymmetric decay reported from BaBar. Why, because the lab recording the gap measured from Supernova 1987 was not "expecting" to observe the data obtained. Therefore, there were no preconceptions or expectations for the measurement that was obtained. This is not true for the BaBar facility. The whole purpose of this experiment was to observe/catalogue proof of asymmetric decay.

No, I am not accusing the folks at BaBar of falsifying data. Rather, I believe that it is most likely that the data from that experiment was misinterpreted based on wishful thinking. Look closely at this experiment and you will find that the data-set of "positive" asymmetric sightings is quite small compared to all events that occurred, and the margins of error are quite high. A Russian scientist with whom I was working (and an expert on detector technology) confided that the results used to "prove" asymmetric decay could even be attributed to a single defective CEM—which if this is the case, then no asymmetric events actually occurred. Assuming, however, that BaBar did measure asymmetric events, another way to account for them is that there is a (yet to be detected) mirror asymmetric event that reestablishes symmetry between matter and antimatter. Regardless, these results are inconclusive and not universally accepted by the scientific community. On the other hand, the data obtained from Supernova 1987 is both conclusive and accepted—even though up until this new model there was no easy way to account for it. Supernova 1987 supplies very strong supporting evidence in favor of the Dominium prediction of opposites-repel interaction between matter and anti-matter. The differential gravitational effects on the traveling particles would be expected to produce a significant gap in the arrivals of the two peaks of the two different particles. This is exactly what was measured.

## Rebuttals to "Common Knowledge"

Unedited posting to a comment-thread January 2008

***History behind this interchange:***

*As mentioned, many of my current and former students have tried to help win this campaign. One former student, Daniel Y, is currently attending Brown University. The notion of the Dominium model and the danger implicated in LHC intrigued him and he questioned the entire physics faculty at Brown. He came across Prof Greg Landsburg, who is an expert in particle physics and who was one of the first people a decade ago to suggest the possibility that LHC could be capable of generating man's first synthetic black-hole. Since that time he has shifted his position and is now a proponent of LHC. He outlined to Daniel three reasons why LHC will be safe, whether or not it does generate a black-hole. Although Prof Landsburg refused to post anything to the SciAm blog (he was much too busy) I encouraged Daniel to post those arguments so that I might respond. Therefore, in response to the first blog article Daniel Y posted the following explanation of the current "Common Knowledge" LHC safety arguments were posted to the first blog article are:*

1: Computer simulations suggest no reason why synthetic black-holes won't revert back to common forms of matter, i.e., samples will be unstable.
2: The experiment will be conducted in a vacuum, far away from the solid machinery; therefore will not be able to accrete (feed).
3: Any produced sample will have very high KE, far more than needed to escape the Earth, therefore, even if stable, samples will escape to outer-space.

Rebuttals:
1) The first point places hope on various "computer simulations" to predict the unknown outcomes of a yet to be conducted experiment, all things being equal, of course. The problem here is that a computer simulation is only as good as the basic parameters defined by the computer programmer. The fact is, this experiment has not been conducted. Any simulation, is just a guess, albeit fancy, high-tech, and glitzy. If you change the initial parameters and assumptions, you can achieve a different outcome. By adjusting assumptions you can obtain vastly contradictory results.

Evidence of the unreliability of "computer simulations" can be found from the fact that there are computer simulations published throughout the scientific literature that portend to "prove" vastly different theories. A true skeptic would ask, "Who commissioned this computer simulation, and why?" This is an important question. Wishful thinking, a desire to appease the public or to justify direction to financial backers could be motivation to choose parameters that favor a particular predetermined outcome. I've heard professors tell their grad-students, "Make me a model that show x, y, and z." If they produce such a model, does that mean that they proved that x and y lead to z? No, all they did was dramatize the fantasy of their professor and complete an assignment.

At the very best, computer simulations approximate nature, they do not substitute actual experimentation. However, in order to closely approximate a situation, one needs to know the exact dynamics of the situation being modeled before beginning. This fact leads in to the next fault: in order to construct a computer simulation of the dynamics of a mini black-hole, one must presuppose understanding of the exact dynamics of the four known forces of nature (gravity, electromagnetic, strong, and weak) under exotic conditions where matter is forced within extremely close proximity. All of these dynamics are poorly understood, because the experiments have yet to be conclusively conducted. Therefore, the modelers are forced to make sizable guesses. Each guess will have a degree of error; each bit of added error will have a multiplying effect; and as a result the outcome produced by the model can be warped 180 degrees away from reality.

Also, to state that there is no reason to believe that the synthetic sample will be stable ignores both this model and previous observations of black-holes. For example, one abandoned model for black-holes was that they are portals for travel between one area of the Universe and another. Although science fiction writers still use this model (because it leads to interesting storylines) it is not accepted by the scientific community. Why? Because if it were true, then scientists should observe matter appearing out of black-holes as well as observations of it entering. However, that is not observed. Recent observations do attribute the generation of high speed cosmic ray particles to black-holes of neighboring galaxies, but that in only in conjunction to ones, which are feeding. These particles most probably emanated from that which was being consumed not from the black-hole itself—much like a peach squirting juices because it is bitten. Black-holes themselves appear extremely stable.

The new Dominium model also provides lines of reason to suggest that mini black-holes will be stable. The analysis that provides explanation for "dark matter" (conclusions 13 to 20) is the main aspect of the model that points toward the stability of mini black-holes. If "dark matter" truly were mini black-holes that were formed in the Dark Event, then they are extremely stable. This conclusion is reached by their apparently inert behavior. We know of the existence of "dark matter" only from its gravitational side-effects. If it were even slightly unstable we would have detected its presence using other means, but we haven't been able.

Points 2 & 3: Let me address these two points together first, because they are contradictory. On one, point 2 suggests that the event will occur in a vacuum far away from any matter that could potentially be consumed. On the other hand, point 3 states that in the worst case condition the created sample will have so much latent velocity that it will escape the Earth's environment. Question: how can both be correct? How can you be contained so deep within a vacuum that you won't be able to contact anything, simultaneously have so much energy that you won't be contained? You can't have it both ways.

Point 2 is correct: initial collisions will occur within a vacuum. However, this does not imply safety, the size of that vacuum is not limitless. True, compared to the size of a proton, the vacuum is "large," yet compared to the size of a human, the diameter of the LHC been is puny. The biggest problem with the second point is the third point: the created samples will possess great velocities. Essentially, point 3 negates point 2. Once a sample of black-hole material is generated, it will have velocity enough to carry it out of the vacuum protected area and crash into the wall of the containment vessel. Once it hits, it will be able to accrete (consume).

Point 3: This seems like a magic escape clause: "even if stable black-hole material is formed it will be moving at speeds higher than those needed to escape the Earth." The problem is, this experiment is not set up like Cape Canaveral, pointed up to clear skies and open space. The LHC machine is deep underground. In order for any sample to escape the Earth, it must first pass through over a 100 meters of solid machinery, rock, and soil. Undoubtedly, along the way it will bump into obstructions. At each collision it will be slowed. At each collision the Beast will be able to accrete (feed). And with each accretion it will gain mass, loose momentum, and be less and less able to exit the Earth. Because of the densities of materials and distances to be passed through, it seems highly unlikely that a generated stable black-hole sample would ever escape the Earth.

Remember also: the previous paragraph describes the best-case scenario if after collision the newly stable black-hole is deflected perfectly upwards. There are an infinite number of other trajectories that the new black-hole could take. All of these other paths involve the necessity of passing through solid material more than 100 meters thick. In some cases, a lot more—the worst case scenario is that the new stable black-hole is directed down and, therefore, must pass through the entire Earth. This is truly a bogus assurance; all things are not equal, the energy needed to launch a particle at Cape Canaveral is not the same for a particle being shot in a random direction through solid rock.

Another consideration that increases the apparent danger of LHC is point #3, itself: because the newly formed black-hole material will be shot through solid matter, collisions will necessarily occur. Because of this, with each collision the newly formed Beast will be force-fed more and more matter. With each feeding, its stability would be expected to increase, as would its own ability to feed itself. Instead of dispelling the danger of LHC, point #3 highlights reason for concern.[4]

---

[4] Prof Landsburg apparently would not discuss any of the rebuttals.

**Coincidence, Imagination, and/or Providence – You Decide.**

Unedited and last posted article to Scientific American, May 2008

Who is Hasanuddin? How could a new model appear at this very moment—one that points to grave dangers of an experimental machine, LHC, a hair's breath away from it being turned on? Do we, as individuals, matter at all?

The journey of the last year has been the wildest, most arduous, and most odd of my entire life. There are no adjectives to describe it. Let me start off by saying, "I don't know why any of this has happened, but it has happened." Today, there is a beautiful new seamless model that answers most of the modern unanswered questions of astrophysics. The model begins, however, at a very basal particle physics question—how does matter and antimatter fundamentally interact & what happened to all the antimatter that should have been produced during the last Big Bang? The methodology of the new model is both revolutionary and archaic—the formal Aristotelian syllogism. Though this methodology served to create the original foundations of all scientific disciplines, over time new applications of the deductive syllogism became increasingly scarce, and, eventually, it fell into obscurity and was forgotten by the bulk of the scientific community. The model was published originally in November 2007. Since then, the abridged version has been downloaded several thousand times from http://www.sendspace.com/pro/dl/u56srb It has been attacked and assailed on this blog for the past five months. To date, none of the detractors have found a single flaw within its construct or the scientific premises used. Ironically, three of the wanna-be detractors ended up supplying new lines of supporting evidence that were not used in the original book. The imperviousness of the new model is even more remarkable when one considers that there is a huge motivating factor that should push many to want to overturn it—its ominous implication that the very expensive and massive mach-ine, LHC, could "succeed" in exterminating all Life, from man to microbe. It is an amazing coincidence that this model should appear now, just moments before this machine is to be turned on. However, the manner of how this new model came into being is almost more unbelievable than the fact that it exists at all.

Who are we? Just tiny grains of sand of an infinite beach. We all eat, drink, and use the toilet—we live, and we die. So. Of all questions, this is the biggest and most daunting.

Who am I? I am just another grain of sand—that is the easiest answer. Anything more specific falls short of the truth. Many different people have called me an enigma; I don't try to be. I am an outsider; again, I don't try to be, I just am. Ever since being the first white-boy to attend SMA Negeri 1 Ujung Pandang, I have been an outsider. Though I love people and society, I have never quite fit in perfectly. But, I'm okay with that. I question things that others accept as standard truths—anyone who actively questions social norms can end up as an outsider. I suppose my questioning of societal norms also began with my cross-cultural emersion.

Though (coincidentally) I am the son of a Presbyterian minister and was well taught in Sunday school, I even actively questioned my Faith. What choice did I have in being born into that family? Couldn't I just as easily been born the son of a Rabbi or tribesman? I converted because I saw an entire city functioning very well using the understandings of one religion, Islam—though I do not know with certainty if even my chosen path is "most correct," but it works for me. I also converted because it was the first religion that I ever encountered that, in writing, states that Jews, Christians, and all others have a chance to reach Heaven depend-ding on how they lived their lives. That paradox had always been my biggest sticking point with Christianity: if the only way to Heaven was through Jesus Christ, then how could a "loving God" damn the Chinese millions who have the "misfortune" of being born into a non-Christian home? Yes, I've heard people try to rationalize an answer, however, nowhere in the Bible did I find a written explanation. There were other virtues that caused my conversion, but that's a story for another day and not on this forum. However, one central truth that comes from Islam that I hold on to that is relevant is that I am just a tiny, insig-nificant, hardly noticeable speck of sand. I am just a glimmer in time.

Upon return to California I was an outsider still. People even asked me what country I was from because somehow I picked up an accent. (I still don't know how that is possible because I wasn't even speaking English while I was over there.) Though I have kept my Islamic Faith, I have only been to an American mosque only a handful of times. I know that goes against the "dictates" of the faith, but I am an outsider there too. Honestly, the few times I did go, in the 80's and 90's, I was so turned off by some of the semi-slanderous words being used to describe the evil white-American society that I wouldn't go back.

Also upon returning to the USA, I entered and attended College of Marin and scrambled for odd-jobs. Walking through campus one day I saw an opportunity. There was a flyer offering money for being a tutor. Instantly, I knew the perfect class: Philosophy 112, Introduction to Logic. I excelled in the subject, but most students were struggling and complaining. When I went into the tutoring office they laughed at me. "A tutor for Philosophy? How hard is it to philosophize? Anyone can do that!" I pleaded my case and they agreed that if I could find one tutee they'd sign me on. I did, and soon I had more work than any of the other tutors. Philosophy 112 was a class that instructed the fundamentals of classical formal Aristotelian logic. I was under near constant employment for the next two years.

While at College of Marin I also took and excelled in every science course that was offered, even Microbiology for RN's. I was in no rush to get out of school; I loved the opportunity to learn and to question. I was awarded a Rotary ambassa-dorial scholarship to study one year abroad at the Universiti Pertanian Malaysia. I studied Tropical Agriculture for one year and continued my understanding of Islam. I entered my bachelors with over one hundred units of excess science course-work. UCDavis I graduated in only 1.66 years—I was paying for it myself, so I doubled up my classes to reduce costs. My "worst" term I carried 29 hard-core science credits, where the school-recommended load was 12.

The coincidence of graduating with my masters right in the middle of the recession forced me to find the odd-job of substitute teaching in a rough urban high school, just so that I could afford to eat. Although the initiation can only be described as pure living Hell, the job fit me and I was good at it. Besides, I am just a grain of sand—I have never had aspirations to have anything more than a quiet life. The vocation of teaching also fit perfectly with my ideals (again a long story for another time and place.) The point is, since then, I have never strived to be anything more. There is no way I would even considered "climbing the ladder" to become a school administrator. What I do strive for is to be the very best teacher possible and to continue learning. I attended conferences, signed up for lectures, read the current literature, and attended every science-oriented learning opportunity that crossed my path. It is in these continuing education workshops, where I became aware of and interested in the fringes of understand-ing in the fields of particle and astrophysics. Coincidentally, all of the major workshops that I attended have now come in convergence to create the new Dominium model:

1: In 2000 I was chosen by NSF to represent the United States at the HST program at CERN. The program was an intense seven-week regiment. I attended the student summer lecture series given by Nobel laureates and top scientists; I went on lots of tours of working labs and interviewed their technicians; and I worked in a CEM lab.

2: In 2003 (or maybe 2002) I was chosen by NASA to represent Massachusetts at the NEW program at Goddard SFC. This was another (less intense) three-week program, which focused on astrophysics.

3: In 2007, I was chosen by the MIT Alumni Assn to represent Boston, and attend the one-week SEPT lecture series and program. It is here where the current story begins.

In a lecture titled, "Physics Frontiers," the lecturer made an incredibly stupid misstatement. He said, "Now that we understand the dynamics of the Higgs boson..." My hand went up in a flash. He was obviously surprised that anyone would have a question at this point in his lecture, but I asked anyway, "Sir, how can you say that we understand the dynamics of a particle that has never been detected nor has any evidence been produced that this particle even exists." He backpedaled and stammered. Finally he did have to admit my point—no, there is no actual physical evidence supporting the theory of the Higgs boson, only fancy mathematical constructs—but he was confident that it was just a matter of time. After the lecture, I went down to talk to him. I asked him what his thoughts were on an alternate model to explain the Big Bang that I had first learned about at CERN and had been mulling over ever since. To my surprise, the MIT professor had never even heard of that alternate model.

I went home and began to type a letter to that professor. I wanted to explain to him the model that seemed to hold great promise. Though I had first learned about it seven years prior, many things happened in the interim (again, another story) that had added new bits of insight. As soon as my letter started, I knew that the issues were so complicated that I needed a way to formally tie everything down and organize. I used a syllogistic method, because to me, it is the most succinct, clear, and mathematical way of presenting complex components of a giant system into order. The syllogism grew amazingly fast. I couldn't type as fast as the connections were being made. It didn't take me very long at all to realize that something new was being fashioned.

That was one of the weirdest weeks of my life. I continued attending the MIT lecture series, but I would race home to type. Coincidence: every single time I descended into the subway that week (12 times in a row) an Ashmont train was either at the station or just entering. Coincidence: every receipt I got a store receipt that week somehow the numbers all lined up (for example, $22.22, $9.87, $18.18, $20.00, $44.44, etc.) Coincidence: I did skip a few lectures so that I could type, one class I somehow "knew" I couldn't skip—Nanotechnology Frontiers—from that class I learned about "Self-assembly" which turns out to be the missing link that makes this model complete. Coincidence: this all occurred at the very beginning of summer vacation and I had all summer to write, which I did.

The focus switched, I knew this model was something new. Therefore it must be researched, referenced, and written in a way for it to be accepted into journal publication. Coincidence, I live within walking distance of the UMASS library which is completely equipped to research all modern scientific journals. I spent the summer in that library finding references from the current literature that backed up the premises and assumptions being made by the growing model. In the mean time, I continued to write.

While writing two of the strangest illogical emotions came welling up inside of me. One emotion was an exuberate feeling of joy, wonder, and realization that this may be the dawning of a new age of science and prosperity. The other welling feeling was the polar opposite, an almost suffocating feeling of doom, dread, death, and horror. I was two weeks into writing the model when the next step in constructing the model pointed deductively that mini black-holes will be stable. Instantly, memories flooded back of being in a CERN lecture-hall hearing the speaker talk triumphantly that, the then proposed, LHC would win the world-wide race among labs to be the first to synthesize man's first sample of black-hole material. If that happens, and if the model is correct, then the source of the well-ing feelings of danger was explained. Again, it is coincidental that those illogical feelings of dread preceded the identification of potential danger.

The direction of this work changed again. Because of the danger, haste was called for. This model had to be made known ASAP if there was to be any hope in it having any affect in stopping the apparently ill-advised LHC project. Coincidence: I'm married to a graphic designer so cover design was easy. Coincidence: among our best friends is the entire production staff at MIT Press so I was able to have it professionally edited and coached on how to get this thing out.

Immediately before the cover was to be submitted a child came into by room and shoved something in my hand and said, "Hey look at this, it's cool," he said, and then turned and walked out. Honestly, I can't remember who the child was, it happened so quickly. None of my current students admit to being the one who gave it to me. On the slip of paper was a word tree that amazingly, and beautifully, fit the crisis at hand. As soon as I saw it I couldn't see anything but how it correlates to the Earth being consumed by a black-hole. Coincidence, yes, and I incorporated it into the cover design. For those who don't have the book I will include it here (notice how every word coincidentally and ominously aligns.)

S T A R T L I N G
S T AR T I N G
S T A R I N G
S T R I N G
S T I N G
S I N G
S I N
I N
I

The most bizarre, and ominous, of the coincidences are those that align with either Biblical or Quranic narratives. As discussed many times on this blog, the predicted demise of this planet after the creation of stable matter-based mini black-hole material tightly parallels the stories of the apocalypse: fire, brimstone, sulfurous gases, and everyone being sucked down to Hell. What I did not ever mention is the other links to the book of Revelations.

From my very first blog article, I did allude to these coincidental parallels. For example, I have consistently referred to the potential synthetic black-hole as a tiny "Beast." My choice of words was no accident. I never explicitly explained my-self because I wanted debate to focus on science, not the interpretation of scripture or these bizarre coincidences. I know that many scientists are in-your-face athe-ists. I knew that to mention my religious beliefs or these coincidences would be to cause such people to go crazed. Besides, I did not want the model to lose credibil-lity before it had the chance of being scientifically evaluated. But there is more.

One coincidence that I found extremely unnerving has to do with the logo used by CERN itself. Actually, it is a beautiful logo that cleverly incorporates the fundamental design of a loop-accelerator in an overlapping pattern. The heart of a loop accelerator looks like a "6." The circle represents the main loop accelerator and tangentially attached is a linear injection accelerator. Check out the CERN logo yourself. How many sixes do you see? I see exactly three true sixes and two figures that are the reverse of a six—coincidence. Please don't misread this observation and confuse the Bible with Hollywood. Nowhere in the reading of Revelations do I see a connection between the "Beast" and anything evil, like Satan. Rather, the Beast is connected to the End of the Earth only. Though that might seem like a "bad/evil" thing, it is actually just a neutral event. I mean no disrespect against CERN, its employees, or its past contributions to society, nor do I think that there is anything inherently evil about that facility. All I'm pointing out is an ominous graphic design coincidence—a coincidence that I never observed until this this August. This observation occurred by being woke at 4am with the strangest welling feeling that I must to look at the CERN logo again (the same way the bulk of these writings were inspired.) Coincidentally, I knew what I was going to see before my eyes actually confirmed it—three, and only three, true sixes. Now, read the book of Revelations. Within the passages relating to the Beast, what does the Bible instruct: "Beware men who calculate the number of the Beast." In that line lays another coincidence. Who are the new Dominium model's most ardent detractors? Scientists who have defined their positions based on "calculations"—not physical evidence. The connection between the Domin-ium model and LHC boils down to the Dominium model versus the notion of Hawking Radiation. Proponents of the machine site the evaporating-assurances of this theory to allege that there is no risk posed by LHC. However, there is absolutely no tangible documented physical evidence to support the notion of Hawking Radiation—just lots of fancy-math "calculations."

One coincidental connection, that even I was floored to see, comes from the Revelations passage of the horsemen of the apocalypse. Shortly after my first posted article appeared on Scientific American, another SciAm blogger, Quantum Poetics, posted this little ditty:

<u>The Four Horsemen of Physics</u>
Horsepower is The Four Horsemen that are The Four Forces that will come together at the end of the world: conquest (gravity) rides a white horse; war (electromagnetic) a red horse; famine (strong force) a black horse; plague (weak force) a pale horse.

Replace the first-word "Horsepower" with "Black-hole formation" and you get a very powerful omen. The idea that all four fundamental forces will come together and work in concert during black-hole formation has always seemed to be a basic assumption to me. However, to see someone make the parallel between the four fundamental forces and the horsemen of the apocalypse was unexpected and unnerving. Coincidence.

Consider next the "ray" of Solomon projected to reign death and destruction, which is also predicted within Revelations. What is LHC but a machine that produces two extremely powerful rays of particles and causes them to impact head-on against each other? Coincidence.

Even my own birth seemed to be marred with coincidence. At age 1, my would-be older sister Jane came down with infant leukemia. My mother has always been certain that she was a fatality of the radioactive fallout exposure endured by populations downwind of aboveground atomic test-sites. The increase in infant leukemia downwind of all aboveground test-sites was a major determinant of the current world-wide moratorium against further aboveground testing. Jane died in February 1964; I was born in February 1965. My mother was undoubtedly grieving during the entire pregnancy. Coincidence: I was born as a direct result of massive scientific projects that failed to contain synthetic contaminants that lead to the deaths of innocents; today I fight to preempt a project that would kill all Life if it failed to contain its synthetic creations.

Honestly, when do the number of coincidences cease being coincidental? I am a very grounded and practical man. I question standard assumptions as part of my inherent make-up. I am intensely skeptical. Yet, given the number of coincidences, even I am swayed into submission. Even my place of residence contains coincidence: I live in a town also known as "Dot" in a state also known as "Mass," how ironic is it that I live and work in Dot, Mass and I'm fighting to protect us all from being sucked into a dot of mass?

Perhaps by posting this article I will lose what little credibility I have. I can imagine the gleeful attacks of detractors, "A high-school teacher telling Nobel laureates that they are wrong—preposterous!" I am just a grain of sand. Truth be told, so are those Nobel laureates, no matter how they've been superficially elevated by other men.

There is only one month left to go, before the start-up of LHC. Therefore, I have nothing to lose by making known the coincidences that I have observed over the past year. Actually, I haven't mention all of the coincidences that have occurred because it would take to long. Hopefully the reader can also see that the number of coincidences is too large to be coincidental. Though I may be giving fodder to those who would wish to undermine me and the new model, if these "coincidences" are in fact omens, then it would be irresponsible for me not to share them. The fact is, we have one month to go. Therefore, people should know all that I have come to know with respect to the current crisis.

I expect the comments to this thread to be nasty. Oh well, I have divulged much that could be used for rhetorical ad-Hominum attacks. I hope the scientists among the readership stay impartial and stick to the science and logic of the model. None of the coincidences listed in this article are actually connected to the new model itself. To attempt to draw a connection between my mental health and the deducttive syllogistic new model is a fallacious move. The model and the author are independent of one another. Finding fault with me does not address any of the merits of the model—though it might make for good "spin." But, is science about "spinning" data to justify funding or continuation of a pet project, or is science about humanity's quest to understand the truth of nature so that we might better understand our own place within the greater scheme of things? We are all grains of sand. No amount of education, title, or socially imposed stature can change this fact.

Metaphysically, I have no concern which way the chips fall. The fate of the future is now almost completely out of my hands. What decisions are to be made is a function of how the other grains of sand within society choose to react. Yes, I will continue to fight to have this model considered, evaluated, and applied to future and/or current decision-making—until we pass the point of no-return. But, ultimately, the final decision is in the hands of the collective society of the entire planet. Yes CERN was chartered to be autonomous and self-regulating, but the societal collective does possess the power to trump this manmade design. Even if CERN officials do decide to blithely (and arrogantly) go forward, they are still not completely autonomous. There are several scenarios that could stop the project without CERN officials' approval. One hopeful scenario is that those in charge of the CERN's power-supply decide to turn off the electricity to force an open and impartial full reevaluate the entire project. Turning of the electricity for a day right now could set back the project by weeks. Perhaps intervention will come through the politicians, or somewhere else. Time is running out for discussion to occur. But realistically, what is the rush? As a species we have survived for millennia without knowing the answers to any of the questions that could be answered by this massive (and potentially misguided) project. Isn't the rational approach to pause and assess any and all theories that highlight a potential risk? I hope that I am wrong. However, the fact that the Dominium model has withstood five month of critique to come out the other side both impervious and bolstered by the input of wanna-be detractors leads to the conclusion that the model is not flaw-ed. Regardless of the fate of LHC, I will eventually die. That outcome is a fore-gone conclusion. However death comes, I am at peace with my life, including this particular chapter. True, my driving motivation has been the thought of all the students I ever taught dying prematurely—but, no matter what, I could not have saved them from death itself—they also must eventually die, regardless of LHC.

None of the previous discussions explains the first welling illogical emotion that I felt welling through me as I wrote this model—the feeling of joy and of being of the edge of a new era of scientific advancement and reconciliation between all tribes of Earth. Perhaps I can offer a guess. The best answer I can deduce is that those feelings are connected to the possible alternate path of postponing and retooling LHC. If that machine can produce matter-based black-holes (leading to apocalypse) then that machine could also produce antimatter-based black-holes (leading to a plethora of yet unknown applications and uses for the greater good.) But that is just a guess. What is clear is that collectively we are at a crossroads. Our current path could clearly lead to destruction; the alternate path is less clear, but does not lead to a foreseeable doom. Perhaps also, the feeling of anticipated joy comes from the second book, which miraculously, and coincidentally, was also written this summer in the midst of writing and researching the scientific model. The second book deals only with metaphysics, living one's life, and interaction within society. That book will not be published if the course of LHC is not altered—there would be no point. Besides, the decision not to publish came through the same sort of 4am wake-up welling feelings that inspired 90% of all of these writings. I've given up trying to figure the logic of these feelings—I'm just following inspiration. Whether we reach the state of anticipated joy or fall into oblivion is not yet determined. The words of Heinrich Heine eloquently sum up the point that is being made:

**"Be entirely tolerant or not at all; follow the good path or the evil one. To stand at the crossroads requires more strength than you possess."**

Do we matter as individuals? Clearly, the answer is, "Yes." Yet, we are all just insignificant grains of sand. Although this is true, we all count. Just as bags of sand can divert the flow of a mighty river, if enough of us chose to buffet the status-quo stream, LHC can be diverted. But there is very little time left to act. If the coincidences surrounding this model are, in fact, omens, then the actions of those who push for the alternate path will be rewarded in the hereafter, regardless of the outcome. Conversely, those who continue to "calculate the number of the Beast" and/or blindly assail the truths contained within the new model would be expected to be punished for their reckless damnation of future Life and our planet. For those who chose inaction and apathy—I don't know. But since apathy favors the status quo, knowing about this model and choosing not to act seems metaphysically unwise.

Yes, we are all grains of sand. Our lives are totally insignificant individually—even those of Nobel-winners—no-one's life has any more meaning than any other. What we each choose to do with our lives is what gives us significance. Do we use our grain of life to build a better world for those to come? Or do we caste our lives meaninglessly into the black-hole of time? You decide for yourself. Actually, this is a question not exclusive to the dilemma at hand; you decide your answer to this question with every action you've ever taken. I wish everyone clarity with respect to the matters in front of us. Act.[5]

---

[5] Another coincidence that was not included in this article originally, but pointed out by others later, is the remarkable ending of the Mayan calendar in 2012. Like all the other coincidences, this may mean nothing. However, given the worst-case scenario, the current speed at which LHC is being readied, and the expected lagtime between the birth of the Beast and the Hell-like conditions on the planet's surface ...this is a quite extraordinary omen.

## Appendix 2: SciAm discussion with Mr Sheepish—The Man from Tevatron

Because of issues of copyright, the following (unfortunately) will be only one side of a very pivotal conversation on the SciAm blog. The other blogger was a man going by the screen-name, Mr Sheepish. He claimed (and I have no reason to doubt) to work at Tevatron, the largest operating supercollider housed at FermiLab in Illinois. Of all the conversations that took place on the blog, this was the most even-keeled, scientific, and informed. Ideally his words should be included here. Fortunately, I took pains in repeating all the arguments that this man said as part of my rebuttals. Although the conversation lasted over three months (which in and of itself is testimony to the strength of the arguments of this model), this appendix will only contain the portion that transpired over the last month right before SciAm pulled the plug on the blogs. Ironically, one week before the blogs went down, the man from Tevatron supplied me with what appears to be the needed tangible conclusive evidence sealing the case in favor of the validity of the new Dominium model. This text remains unedited from what was originally posted to preserve its integrity.

Good Morning MrSheepish,

Glad to hear weren't intending to be passive aggressive. There really is no need to use normative words when they aren't necessary. With regards to couching descriptions in terms of the proto-Milky Way, I disagree with you when you seem to suggest that this is not advisable because our galaxy is currently a cold place. This galaxy, like all others, evolved from the Big Bang fireball. There is no valid reason why discussions can't be framed around resulting evidence that we are all quite familiar with. At one time, all of the conditions that were in our currently discussion would have applied to the young Milky Way. We all "know" how the story eventually unfolds for the Milky Way, correct? Therefore, framing our discussions around the early moments of this known structure will enable us to more easily see if one of our lines of argument goes astray. You see, if one of us starts drawing conclusions that will not result in producing something similar to the known structure of our cold galaxy, and then we can easily identify that error/misstep. Again, and therefore, please couch all future discussion in terms of the proto-Milky Way, i.e., antimatter-based micelles surrounded by a matter-based matrix.

Summarization of our positions concerning: "3) As a result, the universe clumped into areas that were dominated by matter, and other areas that were dominated by antimatter."

--We both agree that this seems to follow from the first two-step

--We both agree that there is a very short window of time for this to occur

--We both agree that this would occur while the matrix was very hot and as solitary quarks/antiquarks

--We both agree that in order for this to occur, gravitational interactions would have been at an extremely heightened level

Missed points that you did not consider in your analysis of Timing:

**A4: CP Violation weaknesses vs. Dominium strengths**

In your response to A1, you ignore the inherent weaknesses of the CP Violation hypothesis. You indicate that laboratory results suggest that it occurs rapidly. Cool, but you did not mention that CP Violation couldn't account for all the mass that currently appears in the cosmos, even under the most liberal sets of estimations. You ignore the inherent paradox of CP Violation leading to an all matter universe, where all parts should be expected to be gravitationally attracted to one another, yet that this is 100% in conflict with the observation that the universe is still increasing in its rate of expansion. And you ignore the fact that the consensus theory of CP Violation is in direct contradiction to the Standard Model's assumption of charge, parity, and time SYMMETRY. Therefore, to defend CP Violation you are standing against the Standard Model. That is a large number of paradoxes, incongruities, and conflicts to swallow all at once. By ignoring all of these things, essentially that is what you have done.

Also, by ignoring all of these issues, you also ignore that the juxtaposed Dominium model does not possess any of these incongruities, paradoxes, or conflicts. There is no paradox between the model observations of a Universe increasingly expanding at higher rates of acceleration, because the alternating supermatrix of dominia would be mutually gravitationally repelling its neighbors. The is no conflict with the Standard Model, because the new model does not insist on an all-matter Universe. In fact, the new Dominium model 100% supports the Standard Model's assumption of CPT symmetry.

This is a lot to selectively ignore

**A5 (Also "G" but it would directly affect timing): Gravitational primary stability (self-sorting) as the primary driver of the initial Big Bang fireball.**

You did not comment on the notion of gravitational primary stability, i.e., self-sorting being a primary driver of this exotic and extremely heterogeneous initial system. It is surprising that you did not comment on this possibility because it provides both reason to answer all of the above criteria for the Dark Event that we both seem to agree upon. If gravitational primary stability, i.e., sorted configuration (immiscibility), was the primary driver of the system, then the extreme speed, shortness of time needed, and level of ballooning are all given reason. This is not a trivial consideration.

**A6: Dominium Model's Premise #3: Ballooning affects at extremely close ranges**

You put forward that you are skeptical about timing because in order for the Dark Event to occur, that ballooning of the gravitational forces felt had to occur. Your exact words are, "this depends on how strong gravity gets at close range. If it inflates in strength many orders of magnitude it might begin to approach the strength of the EM interactions you describe." Your statement seems as a qualification that you "came up with" yourself. The fact is we are in agreement on this matter and it is discussed within the book. Premise #3 of the model makes this statement very clearly. This is one of the very few premises that are used by this model. Are you simply ignoring that this premise was delineated or are you questioning whether the magnitude of ballooning could be as high as would be necessary to achieve the Dark Event? I believe that we are in agreement that given enough of a ballooning of force at close proximity.

Given the points of this discussion that you ignored, I can see how you might feel like you don't have enough to make a conclusion. When you add in the points that you ignored it becomes easier to accept that the Dark Event could indeed occur in the first fractions of a second after Big Bang,
-- A4: CP Violation weaknesses vs. Dominium strengths
-- A5: (Also "G" but noted here because it would directly affect timing): Gravitational primary stability (self-sorting) as the primary driver
-- A6: Dominium Model's Premise #3: Ballooning affects at extremely close ranges then it becomes easy to facilitate a conclusion: The Dark Event is plausible, follows from the central hypothesizes & premises, and coincides with plenty of documented verifiable observations.

and also, the to be introduced points of this post

-- A7: Possible aid of Strong and Weak forces to facilitate the Timing/occuarnce of the Dark Event
-- A8: Hyperinflation period predicted by Hubble acceleration graph shows particles moving faster than the speed of light

Now on to your point #8 (Adiabatic Expansion) where you voice skepticism in adiabatic expansion of remained mass of the proto-Milky Way (i.e., matter for our embryonic galaxy) that was not collapsed into either antimatter micellular black-holes (AMBH) or the proto supermassive galactic central black-hole (SGCBH)

**B1: Was adiabatic expansion aided by the receding seam "dragging" the uncollapsed opposite type material into the volume of space the forming MBH once occupied?**

This is how this discussion was actually originally entitled. Somehow, you gave it a different title, which is way off from what I described, "Did effects from general relativity cause a drag bringing in the dominia material during the collapse?" Then you go on to admit that you're a star with general relativity. Thanks for sharing, but general relativity was not what I was trying to describe in my original post.

I originally said, "The argument for being dragged in comes from my understanding of the current theoretical and observational research on the matter of black-holes. (Excuse the fact that I am very bad with names and have forgotten the authorship of the papers I am referring to.) However, I have read multiple papers that describe a boundary (or seam) between regions of normal space and of a black-hole. In some theories of black-hole formation, this seam is knit into the fabric of space-time of the "normal" regions. From that premise, the following syllogism flows:

--If it is true, that space-kit knits itself to the seam of a black-hole
--Then when the seam was first formed there was a degree of space-time knitting,
--If the seam moved through space as the MBH collapsed/formed,
--Then opposite-type material would be dragged along also"

Sorry if the reference to space-time made you feel like I was speaking relativisticly. It is not the concept of space-time that is important in this discuss-sion, but the idea of a moving boundary that leads to the posited adiabatic expansion. For example, past laboratory thermodynamic experiments to explore the adiabatic phenomenon involved moving the boundaries of some enclosed volume of gas. The boundaries in this case could be the surface boundary between the gas of the experiment and the steel of the experimental vessel. For the case of the Dark Event, the moving boundary would be between regions of space occupied by matter versus regions of space occupied by antimatter.
--As black-hole material is formed, one aspect that we know for sure, is that the volume it will occupy will decrease by 99.99999% (etc w/ the 9's) less than it originally was.
--Therefore, as the black-hole was formed, the boundary delineating the immiscible border between the antimatter micelle forming the AMBH and the unincorporated matter surrounding it would move to adjust to the shrinking new volumes of the micelle, until the process is complete.
--Just as quickly moving steel boundaries induced adiabatic conditions experimentally, the moving immiscible boundaries of the micelles in-process of becoming AMBH within the proto Milky Way would cause adiabatic expansion of the remaining mass of the proto galaxy.

**B2: Uncollapsed opposite-type material would enter the space occupied by the micelle because it is now available**

To this discussion you answer as if you had never read any of the model, "Why? What force was preventing it from entering the space before that is now absent? I can't enter a solid wall of rock because of electromagnetic repulsion between the rock and my body. If I were made of antimatter and the rock were matter, however, then I would not be prevented from entering, rather matter-antimatter annihilation would occur resulting in a huge explosion."

The analogy of you entering a rock is quite appropriate. Notice that your example (rock/flesh) is quite similar to the experimental design example I used in B1 (exp.gas/exp.steel container.) The last sentence, however, is wild speculation based on no evidence, relation to the model, nor observed confirmation. The fact is, we have never created conditions where we manipulated a large amount of antimatter against an equally large amount of matter. All past experiments are highly skewed in favor of matter, because all the machinery and surrounding materials are all matter based and the quantities of antimatter were all puny in comparison. Therefore, to say that you could "easily" enter a region densely populated solely of antimatter is shear fantasy and speculation on your part.

As for your initial question in this reply, "What force was preventing it from entering the space"—well, that would be the gravitational repulsion (& the interactions of the other three fundamental forces) that were responsible to the immiscible boundary in the first place.

**A7: Possible aid of Strong and Weak forces to facilitate the Timing and/or occurrence of the Dark Event**

By the way, the exact roles of the strong, weak, and electromagnetic forces are not flushed out by this syllogism or discussion. To say, "all things being equal, they don't matter," would be both stupid and nearsighted. Because protons at the fundamental level are three quarks and antiprotons are three antiquarks, the interacttions of the strong and weak nuclear forces at the initial moments of the Big Bang could have also been instrumental in supplying conditions favoring self-sorting into regions primarily of matter (quarks) versus antimatter (antiquarks.) Such a consideration is supported by the fact that protons do end up with 3 quarks & antiprotons having 3 antiquarks…in other words the known creation of both protons and antiprotons illustrates known examples where sorting MUST have occurred in order to produce the observed results. Therefore, the forces most associated with quark behavior/dynamic could have been extremely involved in the initial process of self-assembly.

**B3: The Hubble expansion graph itself**

You contradict yourself here. First you say there is no data to suggest and inflection point, then you state that the reason for the inflection point is to match the observed CMB data. That's quite a switcher-roo. The fact is, the idea of an inflection point is based on observed data.

**A8: Hyperinflation period predicted by Hubble acceleration graph shows particles moving faster than the speed of light**

You then proceed to correctly acknowledge that during the initial stage (hyperinflation) that in order for this to occur, particles necessarily would have been moving faster than the speed of light. I agree. This observation of yours can be used as further probable cause to alleviate your skepticism regarding the mere fraction of a second needed to set the stage for the Dark Event.

How can you then conclude, "your antigravity repulsion would be unable to break the speed of light barrier?" Why not? This conclusion, however, is not backed by evidence, reasoning, or anything other than opinion. To make such a categorical statement you would need a lot more understanding of the dynamics of all four fundamental forces under the extreme conditions of the initial Big Bang fireball than anyone possesses. Therefore, you cannot present such a statement as anything more than a guess based on opinion.

*Also, you neglect the most important detail of trying to deny the Dominium explanation—you supply no alternate explanation of your own to account for the period of initial hyper-inflation or an alternate mechanism that could produce conditions where particles can move faster than the speed of light. Fine, you don't like the hypothesis of repulsive gravitational forces driving this period...okay; please provide us with an alternate explanation that will describe this period. (You don't have one, do you?)

**B4: Your summary of Dark Event triggered adiabatic expansion**

You state: "I still see no reason to think dominia matter would enter the space liberated by the gravitational collapse (of AMBH & SCGBH) on a large scale."

Again, this is still only based on you opinions. You have supplied no reason to suggest that uncollapsed proto-Milky Way matter would not enter into areas once occupied by the material that collapsed.

**CD-(your step 9) Cosmic Microwave Background data**
It seems like we are both guilty of confusing the other.

**CD1: First light**
I can live with the idea that first-light occurred some time, perhaps years, after the Big Bang, that is still fairly very early on in terms of the Universe's overall age. The exact number given in your Wikipedia reference probably has a range of possible actual values given the inherent error in the estimations and assumptions of WMAP's Brian Abbott. Unfortunately, Dr Abbott never details how his number was achieved. Even back-tracing to Dr Abbott's original paper, it is not made clear how Dr Abbott achieves the number that you are quoting. What he does say, and I agree with is, "As the Universe began to expand, the temperature dropped several thousand Kelvin, allowing protons and electrons to combine to form hydrogen atoms." His next statement, the one you zoned in on, is less convincing because it is not tied to a methodology or scientific phenomenon, "This occurred about 379,000 years after the Big Bang." Cool, but the question must be asked, why/how did that estimate come about and what error within those numbers will lead to what range of possible values for possible occurance of first-light after the Big Bang.

**CD2: "Micelle" vs "Dominium/Dominia" confusion**
I understand where you could make this confusion, both are new concepts

Switching those words around, then you are right, after self-assembly, the entire matrix of the Universe would have been arranged as a supermatrix of alternating types of dominia that mutually gravitationally repel one another thereby causing the Universe to continually accelerate in its expansion

**CD3: Confusion between black-holes versus black holes**
To avoid this confusion I have always used a hypen to describe the object, and why I have avoided using those two words to describe dark-colored empty blotches.

**CD4: Your intended point: that the movement of Dominia away from each other would lead to empty blotches on the WMAP images of CMB**
Well, that is a horse of a different color, isn't it? However, even given the maximum amount of time that you wish to use, and the predictions of the Dominium, I still contend that there is no reason to suggest that there will be blank regions of no CMB measurements. The mistake you are now committing is to view the dominia as discrete static points in space, all-things-being-equal, when they are in fact hugely dynamic and evolving structures that are undergoing several sequential changes that occur between the Dark Event and first-light. Even with a huge amount of time elapsed; these sequential events would contribute to the recorded generally uniform distribution of CMB:
1: As already discussed w/ you, micelles at the periphery of each dominia probably would not be collapse into MBH, rather they would be expelled out of the parent dominium to travel to the nearest adjacent like-type dominium.
2: Most MBH formed would be produced toward the centers of the dominium of its creation. Also, most formed MBH (AMBH for the Milky Way) would be expelled because of local asymmetries of force.

So, again, immediately following the Dark Event, uncollapsed opposite-type material would have been ejected from all dominia. After the completion of the ejection of the majority of this material, MBH, comprised of opposite-type material compared to the host dominium, would be ejected from all dominia. As a result of these two predicted ejections, even though the Universe is expanding, the space between proto-galaxies would become choked with migrating material, rather than with an ever increasing void. Therefore, when the expanse became cool enough to produce light, the migrating material would be expected to be among that which "lights up" to produce the first-light recorded in CMB. The subsequent wave of MBH/AMBH would be expected to affect CMB through gravitational lensing affects causing the resulting image to appear even more uniform.

When you originally made this line of argument, you pointed to the a large estimate before the initiation of first-light to lead to expected dark blotches between proto-galaxies (dominia.) Actually, if you consider the two expected waves of material ejected by all dominia, the longer the amount of time between the Big Bang and first-light, the larger the expected amount of blurring on any/all future images of this phenomenon, as in WMAP.

E: -- covered

**F: LHC concern because of MBH Stability and Randomness**

I agree, if the new model is correct and if LHC succeeds in generating MBH then there is a danger. I am glad that you see this.

For the record, these two lines of concern (F&G) are what implicate danger in the LHC machine, the notion that if LHC produces MBH, it will be stable.

Also for the record, let me repeat that I did not "come up" with the idea that LHC could produce black-hole material. That prospect was made known to me at a CERN lecture, where the CERN scientist lecturer promoted this possibility as a good thing.

**G: Levels of Gravitational Stability.**
This line of reason cannot be swept off this thread same manner as "F," because as previously mentioned that this line would also be instrumental in timing, and was therefore reassigned the numbering "A5."

Though there is the ominous LHC connection to this line: being a point of maximum stability, any produced black-hole creation would not be expected to be undone.

---

Dear MrSheepish,

I just formally wanted to thank you for bringing up the question of the CMB picture, the lack of any empty blotches (near uniform distribution throughout), and the apparent conflict that the Dominium model predicts that by this moment in time, the Universe would have already been discreetly organized into a crystalline-like structure of dominia. I am thanking you, not because your example ends up attacking the model, but because you are one of the first people on these threads who were able to show me something "new" to consider.

**CD4—Cosmic Microwave Background (CMB), the Dominium, and Microbiology**
On first glance, you do appear to be correct. But as I described in the last post, that is only if you view the dominia as static discreet objects. However, the new model does not see them as inert, two waves of particles will be ejected from them and radiate outward in all directions. I believe the process of ejecting opposite-type material from its periphery would be an ongoing process that began as soon as the first boundaries between dominia were established. The first ejection would not expected to contain MBH because conditions at the peripheries of dominia do not favor the creation of MBH. Later, a second wave would be ejected containing micelles collapsed into black-hole—"dark matter"—would be ejected (but we needn't worry about this line because MBH would be so inert that it wouldn't be involved with the production of photons within the CMB.) So, let me set the stage, according to the Dominium model, dominia would have been excreting uncollapsed micelles of opposite-type matter. This excreted material would flow outward, similar to the way that bioproteins flow out of the body of a bacterium. If you had a giant colony of bacteria that was arranged in a crystalline formation,

given time (and you quoted 400k years) then the titer of bioprotein within the matrix would approach uniform.

Now the question must be asked, "What exactly is CMB?" That is actually a very interesting question. In common words, it is a shimmer. When we look at it we are surprised by both its apparent near uniformity and what appear to be pattern within the ripples.

We are pretty certain that this shimmer of CMB was caused by areas of space that had cooled below 3000 Kelvin, thereby allowing free protons to "couple" with free electrons to form the first hydrogen atoms. "Coupling" could be considered as the supersymmetric systems equivalent to the initial gravitational self-assembly of Immiscibility, i.e. achievement of a primary level of stability, in this case the level of stability is for the electromagnetic force. Anyway, we know for certain that when new levels of electromagnetic stability are achieved, photons are generated. Above 3000 Kelvin the desire for electromagnetic stability is overcome by the energy available and the particles remain in the plasma state.

According to the new Dominium model, in sectors populated with antimatter, there would be mirror reactions taking place. Antiprotons would "couple" with positrons (the antimatter equivalent to an electron.) Just as a photon would be generated when creating hydrogen, so too would a photon be generated when creating antihydrogen. From Earth, the photons would be indistinguishable from one another because light is the antiparticle of itself.

The analogy of a bacterial colony over time and the titer distribution of excreted bioproteins shows how ejected material could reach a seemingly even distribution around the young, still tightly packed, dominia.

Consider also the prerequisite of 3000 Kelvin. And remember that the CMB is the first-light; therefore it is the first material within the Universe to reach that "cool" temperature. Therefore, one can conclude that the CMB was formed by the "FIRST" material to cool down to that temperature—this does not mean that all material cooled down to that temperature. Within the new model, the candidate to be the objects most likely to cool down the quickest would be the ejected objects. Imagine again the bacterial colony, if once the titer of bioprotein reached a specific level a color change is induced. If this crystalline-structured colony stretched outward infinitely, and if all the bacterium had been created and positioned at exactly the same moment of time. Then the history of ejected bioprotein would be the same for all. Therefore, the color-change "shimmer" would occur throughout the matrix at nearly the same time.

**CD5—CMB, the Dominium, and expected/observed subtle patterning.**
Another interesting result of this analysis also lines up with observed data. If dominia were sending out waves of ejected particles, then resonance patterns might be expected to be observed within the data. As mentioned before, one of the surprising findings of CMB was the observance of subtle fluctuations that appear

to many to be pattern-like. Resonance patterns would be expected to be clear and distinct if all dominia were exact in size; hence ejection history. However, that is not the case. The Dominium model predicts that the size of dominia will vary therefore their exact positioning within the overall crystalline matrix will vary as will their eject/wave history. Therefore, any expected resonance patterns would be expected to be less distinct and more cacophonous. This expectation matches the observed data.
Posted +/- June 17th

---

Dear MrSheepish,

Sorry I'm not waiting for a response, but this topic is too interesting and my own last post carries repercussions that should be responded to and clarified. With regard to the discussions of the data of the CMB collection and the Dominium model, there are a few more very interesting things for you to consider.

**CD6 – Problems with the bacterial analogy, hyperinflation**
There are several problems with the bacteria colony producing bioproteins to show how dominia expelling opposite-type micelles of uncollapsed material into the matrix between dominial cores.
1) Bacteria can only be set so close to each other.
2) Hyperinflation will occur for dominia; while bacteria are relatively static.
3) Dominial cores would only account for a tiny fraction of the volume of the sphere at any one moment in time. However, as time progresses, dominial cores would be an exponentially smaller portion of the overall total volume of the super-matrix; while bacteria can account for a huge proportion of the overall matrix.
4) Bacteria divide; if anything, like-like adjacent dominia would be expected to merge.

Knowing the problems with an analogy is one of the best ways of figuring out the true nature of a system being analyzed.

**CD6-1: closeness**
Hopefully this is obvious to everyone. If core understandings of the Big Bang are correct, then all of everything (whatever that may be) was all in one spot at one moment in Time. There is no level of closeness comparable. Therefore, even though we think of galaxies as large and far apart, at one time they were all necessarily much smaller and closer together than bacteria in yogurt.

They would begin excreting opposite-type micelles as soon as immiscible boundaries formed—starting in the hyper-close condition and continuing on through inflation until all internal asymmetrically placed opposite-type micelles are purged for each dominium. And, like bacteria, they would excrete in all directions at once.

**CD6-2: Hyperinflation**

Again a bowl of yogurt and the Big Bang fireball are obviously different in terms of inflation. The bacteria are static, therefore the bioproteins move and are mixed by things like diffusion and the properties of the matrix. However, hyperinflation would effectively smear/mix the micelles being excreted by all dominia, thereby creating an incredibly uniform titer of micelles of all types between dominia. The reasoning is this,

--if hyperinflation is driven by gravitational repulsion
--and if the amount of gravitational repulsion is proportional to mass
--then, more massive objects will feel a greater degree of this repulsion
--therefore dominia would accelerate faster than micelles

*Note, I realize that on first glance this possibility seems to violate Newton's 2nd Law. However, it is known that under these exotic conditions Coulomb's Law alters, and it is a basic assumption of this model that Newton's Universal Gravitation also changes. If the change involves making the relationship between force and mass of a particular object is greater than 1:1. Even a slight deviation would cause the amount of force imparted on a dominium be proportionally stronger than for a micelle, thereby causing the dominia to be accelerated at a greater rate.

Because of hyperinflation, excreted micelles (of all types) would be mixed by the smearing affects of dominia moving past because of greater forces and hence greater rates of acceleration on the extremely mass-dense dominial cores. This hyperinflation-induced smearing/mixing would lead to conditions of uniformity within the intra-core micelle-filled matrix.

**CD6-3: Dominial cores as a percent of the total volume.**

Because of hyperinflation, the significance of dominial cores will decrease exponentially over time. After 397,000 years (or 12.5 trillion seconds) how significant will the dominial cores be in relation to the overall volume? Though this question sounds trivial and unrelated, this is a hugely important question. Even given that there was an inflectional decrease in hyperinflation, the ever-expanding volume of the Universe dilutes the overall dominial core presence exponentially by the second. After 12.5 trillion seconds distances between dominial cores would have become vast.

The importance of this question only truly becomes clear when we look back at the WMAP picture of the CMB. Yes, yes, I know we've already talked about "what" that picture shows (a shimmering that is amazing uniformity with the possibility of subtle patterning)—what we haven't talked about *what* that picture truly is of. That image was taken of photons of particular wavelength all arriving at Earth at the present day, and it represents the furthest back in time that we can currently observe. What this also means is that it represents an image of photons that have been traveling some finite distance dependent on the time they've been traveling. Assuming that all of these photons have been traveling the same amount of time, what the WMAP display of CMB data it a reading of a spherical sampling of objects participating in the shimmer that are all a finite distance away (no further, no closer.) As such, this image is a razor thin sampling of the Universe as it was a finite time ago in these regions that are that finite distance away. We created a map to show this data. That map has area. Now, consider what was stated about dominial core's significance in the overall volume of the universe...what about a cross-section? Through the naked eye, the stars appear to us as a dome of points of light, each the same distance out-of-reach as another. But that is not true. Each planet, star, and galaxy we see in a specific finite distance away. Looking further back into the record, it would be expected that each dominial core also be so finite distance away. For us to perceive a dominial core on the WMAP data of CMB that object must be situated the exact finite distance from Earth that it matches the distance of the shimmer of CMB to reach the Earth. It is, therefore, highly improbable to obtain a cross-section showing the presence of any dominial cores. However, over time watching the CMB, new cross sectional data of the ancient shimmer will arrive at the Earth. As the data continues to compile the chances of encountering a dominial core will rise.

The reason why I bring any of this up is because I believe that if by chance, luck, odds, etc, a single dominial core could show up on the WMAP survey of the CMB, it could help cement this theory once and for all. I believe, it would show up as an empty smudge/blotch with seemingly no emittance. The reason that I make this conclusion is because I would expect the dominial cores to remain above 3000 K longer than the peripheral micelles that were being excreted, simply because of a much more compact nature and probably retain energy. If a dominial core did remain over 3000 K no coupling would yet occur and therefore no photons emitted.

However, on the extremely low chance that a dominial core is caught in the WMAP imagery someday, there are several things that I can used use my understanding of the new model and deduction to predict.

1) If a dominial core was caught in the WMAP data, then it will appear as an empty blotch, as discussed previously.

2) Its "size" would most likely be exaggerated as a result of the inflation making it appear to be much larger as it actually ended up becoming. Just as if there is a small hole in gum that is expanding, by the time it expands, that little imperfection grows amazingly fast. Also, I suspect that the potential for a dominial core itself to expand is nowhere near as great for the dominium at large to expand. Hence, it might "appear" to occupy space many thousands of times larger than predicted for galaxies at those times. (Though there is the open possibility that this hunch is incorrect and the size of the dominium viewed will be preserved in the expanding record and match to the expected size of a galaxy of that age.) *Note also that these assessments are based on a change in perspective. The traditional view of a galaxy is a much smaller entity than what I have defined as a "dominium," hence the need of a new word. A dominium is a noun defining the areas of space, reaching far out, that is controlled by the influence of one dominial core (galaxy

cluster) over another. Boundaries of a dominium are set where influence goes below 50%, and "sovereignty" would switch to a different dominial core.

3) Near to the borders of such a dark blotch, should be a halo of more intense than normal readings. These heightened readings would be a result because that particular dominium would be in the process of excreting opposite-type micelles that would increase titer immediately around that observed core. Its excreted micelles would cause the titer of micelles containing uncollapsed material to be locally heavy. Those liberated micelle would be expected to participate in the CMB shimmer, because they would be now part of the intra-core matrix and therefore be participants in it reaching thermal equilibrium of 3000 K, which allows for coupling and production of the CMB shimmer.

Though I know the chances of catching a dominial core in the WMAP data are quite slim, it is possible—just like the notion of stopping LHC is possible. Because of this possibility, I offer those predictions based on the assumptions of the Dominium model. If in the near future such data is found, then perhaps debate over this model can be settled and quick action is taken to avert potential global disaster at CERN. If someone does find such a match in the data, perhaps the warnings this model brings regarding the stability of mini black-holes will be heeded.

**CD6-4: Galaxies merging – predicted by the Dominium and documented in the evidence**

As stated before on these thread, the new model proposes that the extremely heterogeneous system of the Big Bang fireball was initially driven by an urge to achieve primary gravitational stability (sorting of antimatter away from matter, and v.v.) The combined processes of Immiscibility and self-assembly caused the dominia to form in a more stable (still expanding) crystalline type fashion. Although initial stability had been established, there was variability. Therefore, two like-type infant galaxies could be placed within the supermatrix where mutual attraction could draw them together and eventually merge. This Dominium model conclusion matches reams of data that show galaxies colliding.

By the way, do you remember when the first images conclusively showed two galaxies in the process of colliding came out? Do you remember all the egg on the faces of those who had maintained that—because of previous relativistic and other exotic math ways—galaxies could never collide because of Universal expansion? If you don't remember, I do. Nothing makes me laugh harder than when somebody a little bit too full of themselves has a little air taken out—through evidence that can't be disputed. I had plenty of good laughs during that period. Notice also how quickly those folks modified their equations to fit the evidence, rather than question the worth of assumptions that had led to a faulty conclusion in the first place! (By the way, nothing needed to be modified within the new model to match the data.) In general, people don't like questioning their own basic assumptions. For myself, continually questioning basic assumptions is both interesting and rewarding.

---

Posted on June 18th

---

**The last post to this thread is missing because it was written on the last day that my computer in my classroom before being put away for summer vacation. I was in such a rush that day that I didn't bother to save my words to my disk. Oh well. Perhaps some day Scientific American representative Christie Nichols will stop stonewalling and make good on her promise to release all of the archived discussions to me. Personally, I'd love to read that last post again…it was really quite amazing, even to me.

Essentially the last post was a follow up to the predictions of how the Dominium model would be proven through CMB data that were made in the June 18th post. Basically what I did was run a simple search through scientific journal/news articles using the keywords "hole," "Universe," and "microwave." Almost immediately popped up three different current news articles about a University of Minnisota showing an apparent void spot in the CMB data that conformed exactly to the predictions made above.
1: The area in question appeared as a blotch void of emitted photons (conforming to the prediction that early galactic clusters would have retained heat and therefore been above the 3000 degree Kelvin threshold needed for photon production.
2: The area in question was exaggeratedly large (conforming to the idea that the subsequent rapid expansion of the Universe caused that data set to become distortedly large, much like a pin-prick in an expanding wad of gum.)
3: The void blotch was indeed surrounded by a halo of more intense readings (supporting the idea that the titer of MBH immediately surrounding the galactic core would be higher that the rest of the Universe.)

This story was reported by reuters, space.com, and BBC. Therefore it seems like highly credible evidence!

---

Endnote to Appendix 2
I sincerely apologize that this appendix is not 100% as it appeared online—ideally Mr Sheepish's own word should have been used. I altered nothing however. I have been denied the promised assess to the archived transcripts for over a month. SciAm representative, Christie Nichols, who is/was in charge of the Community forum is not responding to her promise to deliver the actual archived transcripts of those discussions. It is so ironic and regrettable that the decision to erase all proceedings on the Community forum from public view just moments after this concrete evidence came to the forefront after a dialog with a prominent scientist from FermiLab's Tevatron. According to Ms Nichols, the decision was "corporate and non-political." Whatever the true motive, I hope that eventually an edition of this book can be published with the complete transcripts as they actually appeared online.

## Appendix 3: Miscellaneous discussions

This section of this body of work is not going to be structured by a tie-in with a deductive syllogism. The development of the Dominium model, although it primarily occurred during July 2007, has been percolating for over seven years. During the time leading up to July 2007, different aspects of these theories were debated between the author and various people working in the relevant fields. Included in this section is input from experts at MIT, NASA, and CERN were contacted for their input and advice. This section contains arguments used by supporters of older ways of thought (e.g., Einstein) to explain anomalies or counter aspects of the Dominium. Also included in this section are rebuttals to arguments used by detractors who posted to the SciAm blog between Dec 07 to Jul 08.

### The "Inertia Argument" against the Dominium's central hypothesis

As mentioned on page 77, data collected from the phenomenon, known as Supernova 1987, was both anomalous and unexpected. How could neutrinos and anti-neutrinos arrive at the Earth in two distinct peaks? The question whether it is true data is not debatable because this observation was recorded independently at several different labs simultaneously. One hypothesis proposed at the time to explain this data anomaly was, in fact, gravitational repulsion. However, such a possibility would undo standing theories (as previously discussed.) Protectors of the status quo formulae and constructs were outraged and quickly tried to find a flaw with this hypothesis. They united on the "solution" that gravitational repulsion was necessarily flawed because it caused a paradox with inertia…or so they argue.

Their argument is rather simple. If matter and antimatter are opposites, then one of them is the negative of the other. Therefore, they conclude, antimatter would have negative mass. But negative mass is paradoxical to the concept of inertia and kinetic energy. Because of this paradox, hence, gravitational repulsion is a flawed concept.

Although on first glimpse this simplistic argument seems sound, it is not. The flaw comes from the idea that you can apply a negative sign to something simply because it is an opposite. It turns out that it is the arbitrary application of the negative sign that is the flaw (the detractors' move), not the description of the system. True, <u>most</u> cases involving things that are opposite, the mathematical negative sign can be used to form a model to describe that system. However, this devise is not applicable in <u>all</u> cases. There are many opposites that are not the mathematical negative of the other, for example, complementary colors and the sexes.

For this example, the sexes (male vs female) gives the closest analogy to the condition between antimatter and matter. Yes, we call them the opposite sexes, and they are. However, both males and females are equally human. Both males and females possess weight, eat, sleep, live, breathe, and die in roughly the same manner. Just because males are the opposite of females does not mean that it is possible to apply the negative sign to all aspects of males in comparison to females. If one tried, then the result would be paradoxical and nonsensical. For example, if one labeled males as negative and then tried to imply that meant that male weight was negative, then one could make the ridiculous conclusion that the total weight of humanity is close to zero. Effectively that is what proponents of the "Inertia Argument" against gravitational repulsion are doing. When the question of matter and antimatter being opposites arises they arbitrarily assign a negative sign and then proceed to make nonsensical conclu-sions. From the nonsense they claim "victory." Rhetorically clever, but not scientific or logical.

---

**Hawking Radiation**

All due disrespect against the man in the wheelchair, his theory has no verifiable evidence supporting it. Sure, it's got math; it's got symbols; it's got the celebrity status of the most well-known handicapped man behind it – but all of those things are unscientific fluff. What aspect of nature is it tied to? What physical evidence exemplifies it? What tests have been done that show its predictions? The answers to those questions: none, none, and "maybe" LHC.

The fact that Hawking Radiation is the main source of assurance for LHC safety is quite disconcerting. This is a theory that has no physical evidence supporting it. Worse still, this is a theory that hopes to be proven by the very machine whose supporters are using it to justify safety. This is an insane amount of circular reasoning. Safety of the machine sustained by a theory that hopes for its initial verification from that same machine. The 6.6 billion dollar question is what if it is wrong. In that case, we would all pay the price.

Supporters of this theory claim that cosmic ray collisions produce more energy than the collisions that could occur within LHC. Therefore, they conclude, if black-holes can be formed they must evaporate away harmlessly via Hawking Radiation. Although this simplistic reasoning sounds good, one would expect that because cosmic rays are common and much studied, then the observation of a black-hole spontaneously evaporating away should have occurred by now. However, no such data has ever been collected. Also such a comparison is an apples and oranges analogy. Cosmic rays a discrete events, while the collisions within LHC will involve $10^{14}$ protons colliding head on into $10^{14}$ protons. Potentially there is the possibility of dogpiling/snowballing to occur.

The theorists using Hawking Radiation has a safety assurance for LHC are guilty of overlapping use of formal fallacies. The result, something that "sounds" logical, yet is completely unsound. The first fallacy used is *Appeal to Ignorance*: no evidence supports Hawking Radiation and no evidence against, therefore let's accept it as true. The second fallacy used is *Complex Question*: using data from cosmic rays when there is no way to establish a clear link between that natural event and future unknown events at LHC. In this step they are again using the fallacy of *Appeal to Ignorance*, but in a different way this time: no evidence shows cosmic rays forming black-holes, therefore black-holes can't form. To back things up, they use the fallacy of *Bandwagon*: lots of other scientists believe this to be true; therefore it must be true. Implicitly they are also using an *Appeal to Celebrity*: the renowned Stephen Hawking developed this theory; therefore it must be true. And to top things off, they're using *Appeal to Pity*: consider the poor man in the wheelchair, shouldn't he be able to test his unproven theories through the start up their "beautiful" machine? Fallacies are good for rhetoric, but they make for bad science and should never be used for public policy when there is a risk of death of innocent bystanders. In this case, the risk of death of all.

I began this section by stating my disrespect for Mr Hawking. Mistrust is actually a better word. Why should a man confined to a wheelchair for so many years care about the well-being and lives of the able-bodied populations that run, jump, play, and have sex around him? I am serious with this question. Wouldn't, after all these years, he build up a degree of resentment, envy, and animosity towards those healthy crowds that come to "admire" him only to eventually walk away and continue their frolicking? Such dark feelings would almost be natural and expected. The next question is what would a man confined to a wheelchair for all these years care to save from this Earth. He doesn't have any children, does he? As for his own life, what does he truly have to preserve? Honestly, how can anyone trust a man (or his theories) with so little at stake?

---

**Explanation for the absence of antimatter in this corner of the Universe**

Previous theories depended on an asymmetric event to lead to the creation of an entirely matter based universe. On preliminary scan, this theory seems plausible; "no antimatter appeared naturally." However, this statement is made purposefully excluding the huge abundance of positrons within our Sun; after all it was argued, "those positrons are being made through fusion, and you need matter to conduct fusion, so this is a chicken-or-the-egg argument and in this case matter *must* have been first." Since hydrogen matter is needed as a precursor for fusion, and since positrons are a net product of fusion, therefore, the matter-based hydrogen had to be around first—end of discussion. Therefore, the arguments concluded, we live in an entirely matter-based universe—ignoring all the positrons in all the stars.

By making the leap of "ignoring all positrons in all stars," a messy anomaly is discarded. If we were able to ignore all those positrons, then our naturally observed Universe would seem to be devoid of all forms of antimatter. If this were true, then the CPT Violation hypothesis would be strongly supported. However, positrons do exist. Not only do they exist in extremely large quantities, but they are continuously being replenished. To ignore their existence is ludicrous.

As displayed by the Dominium model, the presence of positrons in Helios and other matter-based stars of the Milky Way is not of trivial importance. Positrons, their acceleration by repulsive gravitational forces, and their ultimate escape from the Sun are ultimately responsible for currents within the solar plasma, twisting of the magnetic fields on the solar surface, the acceleration of ions/electrons off the solar surface, and dynamics of the observed solar wind.

---

**Annihilation events**

Matter/antimatter annihilation was thought to be well understood; it was always the final stage of existence of all antimatter created in laboratories. However, the inevitability of this annihilation has led some researchers to falsely conclude that matter and antimatter "seek each other out." Researchers have conducted years' worth of tests involving antimatter. Initially, tests were conducted at high energies, and more recently tests have been performed at lower energies. In all cases examined, annihilation was the ultimate destiny of all antimatter under consideration. Such results could also be used to falsely conclude that antimatter was less stable than matter, thereby supporting CPT Violation theories. From the data, however, neither conclusion necessarily follows. Whether matter and antimatter "seek each other out" or if annihilation is the result of an inherent instability cannot be deduced. It is categorically known that under conditions present in all past laboratory experiments with antimatter, annihilation has always been the ultimate outcome. Exactly how this outcome was reached cannot be determined.

Close examination of the culmination of laboratory tests conducted to date shows that they are all similarly limited in scope and design. All tests are conducted on Earth where amount of matter present dwarfs the amount of antimatter present, which leads to matter to antimatter ratios that are so skewed ($1{:}10^{16}$ or worse) making them inapplicable for use in describing systems closer to the 1:1 ratio predicted by the symmetric creation of antimatter versus matter during the Big Bang, as predicted by *all* models. All experimental tests to date involve the presence of dense solid matter objects in close proximity to the antiparticles. No tests have involved antimatter in either a solid or liquid phase.

Because of the limited scope of laboratory data currently available, it is impossible for any model to categorically propose a mechanism without some degree of error. Even the Dominium model's explanation of laboratory annihilations—the combination of particle kinetic energy and density of other matter being traveled through—is not categorical based on laboratory results alone. What is unique about the Dominium model's explanation of laboratory results is that it is coherent with explanations used to describe phenomena that are anomalies under current theories, such as "dark matter," solar wind, etc.

If the real reason is that all antimatter produced in the lab is produced in a matter-rich environment combined with the fact that the antiparticles in question usually have sizable velocities, this constitutes a concrete framework for the explanation. Unavoidable collisions were actually taking place, because the two prerequisites for miscibility had been met: particle density and kinetic energy. Like the oft-depicted sci-fi fighter racing through an asteroid belt...but this time the asteroid belt is infinitely long. Though the captain tries to avoid the randomly moving rocks, eventually it will happen. Molecules colliding is about the most common event on Earth; the antiparticle had no chance of escaping the lab, let alone this planet. The more traditional explanations—"seeking each other out" or differential stability—though on surface inspection appear to explain the annihilations observed in laboratory tests, do not offer any form of mechanism to justify the assessment. How/why do the matter and antimatter "seek each other out"? How/why does differential stability manifest itself? By not addressing either of these questions concerning the mechanisms involved, proponents of traditional theories gave the impression of explaining laboratory results without actually giving a full answer. Whoever first proposed these partial explanations probably assumed that the mechanical answers would eventually be addressed by the work of those to follow while those who followed could accept the partial explanations as being complete, sufficient, and not worthy of review.

---

**Problems and understandings from LEP**

It is worth noting the evidence collected while CERN's LEP was up and running. At the time LEP was the largest machine ever built. It was designed to accelerate packets (each packet containing billions of individual particles), and electron packets and positron packets were accelerated within a vacuum in opposite directions. These antiparallel beams had been observed to remain within the accelerator for up to 24 hrs before needing to be replenished. This fact supports the Dominium model in the following way: if opposites repel, then when packets of opposite type materials pass through each other, there would be a "steering away" effect between antiparticles in close proximity, and hence a net avoidance effect on collisions. Therefore, the very fact that these two opposing beams could persist for so long is, in fact, evidence in support of the Dominium model. The number of chances they had to collide during those 24 hours is astronomically large—larger that $10^{22}$! If, however, the two antiparallel beams were short lived, this would have supported the other model. Detractors of the Dominium model dismiss such arguments based on statistics. They argue that electrons are extremely small and the width of the particle beam is extremely wide in comparison; therefore, the two particles need to match up exactly in order to produce an annihilation event, and that's why the antiparallel beams can persist and not annihilate whey given over a trillion chances to do so. At first glance, the detractors' argument seems to make sense, electrons and positions are so small and quick that they gave rise to the Heisenberg Uncertainty Principle. However, this argument used by detractors, in fact, supports the new model while offering evidence against older ways of interpreting the world. Buried in their argument is notion that they would "have to line up perfectly." However, the idea that the electron and the positron "have to line up perfectly" is incongruent with another traditional notions concerning anti-matter and matter: that they "seek each other out." If it were true that matter and antimatter "seek each other out," then one would expect that just getting close should suffice in setting up conditions for annihilation. When the effects of elect-trical interactions are considered, the idea that they "have to line up perfectly" makes even less sense. After all, it is well understood that oppositely charged particles attract one another—the positive positron and negative electron should have been attracted to each other through electrical interactions alone. So, why did it take over ten trillion, trillion tries ($10^{22}$) before depleting the packets of electrons and/or positrons while LEP was still running?

The Dominium model offers an explanation that gives reason to the experimental observations seen at LEP. If at the most fundamental level (gravitational), matter and antimatter repel, the statistically large number of missed chances for annihil-lation make sense. This conclusion comes from the same assessment used by detractors that the electron and the positron would "have to line up perfectly." Perfect alignment becomes necessary for annihilation only if at the fundamental level there is an aversion to collision. Therefore, the experimental observation of the longevity of the particle beam inside LEP and the traditional explanations to account for it both offer support to Dominium understandings while undermining older currently assumed models.

---

**LEP "Leakage" Argument**

In the past, some particle physicists discounted the Dominium notion of matter vs. antimatter repulsion by pointing to the observation that leakage of electrons and protons from the beam appears random. "If opposites repel," they maintained, "Wouldn't positrons feel a net gravitational repulsion from the Earth causing them to spill out of the top of the beam? But that is not what we see; leakage appears to be random top versus bottom. Therefore the new model is incorrect."

This reasoning is flawed. Leakage of particles (either positrons or electrons) from LEP has nothing to do with gravitational interactions. Those who present this argument are oversimplifying the dynamics of an accelerator, and they are neglecting the absence of any proof of their own theories: if gravitational interact-tions are always attractive between matter and antimatter, then by the same reasoning, all particles should feel a net gravitational interaction with the Earth and all leakage of all particles should occur out of the bottom of the beam—but that is not what we see either.

Whether one analyzes the leakage of electrons or positrons, the leakage appears random. To account for leakage, one must examine the design of the particle accelerator itself. As mentioned, LEP used strong electric and magnetic fields to accelerate particles. The resulting electric and magnetic forces would supply accelerations many orders of magnitude greater than any expected gravitational interactions, thereby masking any gravitational effects.[Poth] Passing from machine to machine, both packets of particles passed through subtle, currently unavoidable, irregularities in field. As a result, the cross-sectional size and shape of the two beams changed throughout their journey around the ring. This caused particles within individual packets to be in a continuously jumbled state relative to their positions within the packet itself: one moment an individual electron could be at the top of the electron packet, then toward the bottom, the left-center, etc. in a seemingly completely erratic pattern. Because the strong fields used would supply a large force and because particles were continuously jumbled within each packet, all gravitational effects between the beams with respect to the Earth would be masked. Particle leakage can only be concluded to be the effect of random trajectories of individual particles and the irregularities in the man-made accelerating fields. This conclusion supports the observations: random leakage of all particles from the particle beam.

---

**Tying in the Standard Model**

The fact that the classic explanation (Einstein-based mathematical constructs) for the absence of antimatter on Earth relies on an asymmetry to solve the problem is, in itself, hypocritical with respect to the Standard Model. If the fundamental assumption of the Big Bang is symmetry of creation, then the asymmetrical results and conclusions of the CPT Violation hypothesis seems unlikely. The Dominium model solves this issue by maintaining that symmetry still exists, though distribution may appear to be asymmetric at too small a scale of viewing. The assumptions of the Standard Model concerning parity and symmetry hold throughout the Dominium model.

Arguments professing asymmetry, on the other hand, are fundamentally arguments of necessity: "antimatter *must* have disappeared because we see no history of annihilations in the celestial record" or "antimatter *must* have disappeared because it doesn't occur naturally (ignoring positrons in the stars)"... "Therefore," the asymmetric models continue, "the Standard Model must be violated to achieve the current universe." Hence the current theory is known as CPT Violation, because it maintains that parity *must* be violated to achieve the Universe as it currently appears.

As shown through the Dominium explanations, asymmetry is not a requirement to adequately describe the current make-up of the Universe. The absence of annihilation events in the celestial record and the absence of antimatter in this corner of the Universe can both be accounted for without discarding the primary assumptions of the Standard Model.

---

**Anthropomorphic projections**

Inherently, the asymmetric construct of CPT Violation to explain that the entire Universe is now composed of matter has strong anthropomorphic undertones. Listening to someone espouse its virtues can sound like the conversation is actually about sports. Proponents of CPT Violation often come across as biased fans rooting for the home team (matter). "Of course all the antimatter is gone, it was whimpier...good thing, too...life would not get a chance to exist with all those annihilations going on." Deep within this argument is a secondary assessment that the speakers are extrapolating onto themselves. If matter is more virile, durable, tough than antimatter, then so too are we, being made out of the same strong stuff.

The Dominium model offers no such macho satisfaction. Rather, parity is assumed to exist at all levels; neither matter nor antimatter "won" any sort of competition in terms of either decay or annihilation. This would mean that the antimatter never vanished. Rather, symmetry prevails from the beginning and through today. Though our home team, matter, isn't the entire Universe—it would be expected to be exactly 50% of it. To those who may have been concerned about the competitive aspects between matter and antimatter, they can rest satisfied: Where we are, matter is still what matters.

**Qualitative vs. Quantitative**

From the very beginning, this work has been as much about the debate between qualitative analysis versus quantitative methods as it has been between the concept of symmetry and asymmetric solutions to the mystery of the missing antimatter. Detractors of the Dominium model might point to my lack of dependence on math and my heavy dependence on analogy to justify proofs. The approach of this analysis is based on the tools used by the Biologist and Philosopher to define and understand complex systems, i.e., formal logic, experimental evidence, and analogues from nature to add plausibility to conclusions. This approach flies in the face of current methods used to analyze this particular frontier of science--a front-tier begun by Einstein and continued by hundreds of subsequent mathematically driven theoretic physicists. The inherent problem with math-based analyses is that they are not easily "checked" midstream. Another problem with pure math is that it can be used to derive nonsensical answers, like the proof that a former teacher once showed me where one plus one did not equal two. One false assumption, one reversed sign, one missing variable, or one reversed operation will produce a wrong answer. Just because the calculator says "It's so" doesn't necessarily mean you've got the right answer. Yet math students around the world adamantly point to their calculators to justify incorrect answers. The existence of numbers within an explanation tend to give an argument a false sense of validity that seems to surpass the same arguments without the use of numbers. Whether an argument states its issues in terms of "most" or "78.2%" is actually immaterial. Arguments using the percent and decimal could easily be the more flawed in terms of construct, skewed sampling, or faulty alignment of conclusions. Although to human perception, the addition of numbers within an argument adds strength, there is no basis for such an assumption. When one truly assesses all knowledge known to man, the ultimate explanation is written, not numeric. Quantitative proofs without qualitative understand-dings are useless; but conversely qualitative understandings can exist without being quantitatively justified.

For example, in the case of Ptolemy's theory that all heavenly bodies revolved around the Earth, elaborate quantitative proofs existed to explain and predict the paths of planets like Venus and Mars. The elaborate complexity of the mathematical proofs of Ptolemy were "justified" because they were describing the higher order functioning of the heavens (the realm of the gods is necessarily complex). Once Ptolemy's paradigm was replaced by Copernicus's qualitative heliocentric model, more simplistic formulae describing the orbital relationships between the Earth and the other planets followed. Once the true relationships were recognized, it was realized that the true strength of the new model lay in its mathematical simplicity, which had been laid out by a qualitative roadmap. A similar scenario is currently existing: accepted theories are complex, multidimensional, and elaborate to describe the anomalies. The Dominium model is primarily a qualitative assessment that sweeps aside current assumptions and replaces them with a more streamlined construct without anomalies.

**An affront to Einstein**

With this set of theories, part of Einstein's work would be considered flawed, as would all of the work of those who built on flawed mathematical formulae. On the other hand, the Dominium gives new strength and support to the Standard Model, which is something of which Einstein would have surely approved. This model does not depend on identifying where Einstein's mistakes lie; rather the methodology of this approach is to start with a clean slate, make one conjecture, and then follow the chain of expected effects of that conjecture given universally accepted premises that draw on modern discoveries and observations building its support and analogies from nature to supply plausibility. Close attention is paid to the premises used and the way in which conclusions are drawn. It is assumed that all other models should end up with the exact same results if both begin with valid premises and make sound conclusions throughout. Because the Dominium model and CPT Violation do not end up with the same conclusions, they are incompatible with each other, and at least one of them must be flawed. Because the Dominium model is not plagued with multitudes of anomalies, dead ends, and incomplete descriptions that are inherent in current theories, and because the Dominium model is the only model to ever sequence an unbroken chain of events profiling what happens from Big Bang to eventual Big Bang, this model must be considered the more correct of the two. The job of finding Einstein's error is left to the mathematical historians to uncover.

The most likely point at which Einstein's assessment is flawed is that concerning the dynamic of entropy. In light of the recently recorded phenomenon of self-assembly, which seems to be an inherent driving push of primary systems to go from a chaotic and random state to a more ordered design, classic assumptions regarding entropy (all systems seek states of increased chaos) seem wrong. It is classically assumed that disorder is the goal of all systems, but self-assembly of primary mixtures seems to show the opposite: an ordered state is the goal of a heterogeneous primary system, not the expected chaotic state. Therefore, if galactic dominia did become organized via self-assembly to form a crystalline structure it would appear to be defying classic predictions of entropy. Because this defies the most classical assumptions concerning entropy, Einstein could not have employed them in his models; therefore, this is the most likely point at which he made his mistake.

---

**Einstein was a man who lived a long time ago**

The assertion that Einstein could have been wrong is itself a social paradigm challenge. Yes, Einstein was a great man, but the contributions of Newton are arguably much greater and more complete than those of Einstein. But both men, however, were just men, and they both had use for erasers. There are many in society who've elevated Einstein, post-mortem, to godlike status where it is assumed that he was never wrong, absolute in his understanding, and unquestionable in his conclusions and assumptions. Invoking his name is often done to make others cower and stop asking questions: "What are you Einstein?"; "So you think you're smarter than Einstein?"; "Hey look, we got ourselves another Einstein over hear"; or even just "Hey Einstein." I may be wrong with this next statement, but I seem to remember a biographical reference to Einstein fretting that there was a mistake in his design that he hadn't caught. The reason for his concern perhaps came from the fact that symmetry was not forthcoming from his sets of equations. I am not asserting that I am any better/worse than Einstein; both of us are/were just men. Neither of us would be any closer to God/truth than any other human that has ever walked this planet. What makes my particular conclusions here better than Einstein's is that the Dominium model retains symmetry and aligns with what has been observed both in the laboratory and in the sky.

That said, let it be duly noted that I am not discounting all of Einstein's work. The fact that mass can be converted to energy and vice-versa is undisputable. That contribution alone cements his permanent status as one of the great thinkers of all time. In fact, Einstein's theories were used to support conclusions of this paper. The notion that $E=mc^2$ is undisputable and is used throughout this paper. Also, the conceptual model of bending space-time was assimilated into this paper and used as a secondary test of the initial hypothesis.

---

**Biologist's approach to a classic Physics question**

As mentioned previously, a qualitative approach is traditionally used by the biologist to describe complex systems and interactions. Understandings of ecosystems, systematic physiological interactions, and biochemical pathways were all initially understood on a qualitative level long before any quantitative explanation could be developed. Although advancements within the realm of physical science could be described as a progression of theorems and mathematical formulae, which led to increased understanding, it could also be argued that every theorem and formula was actually preceded by a new qualitative description of the phenomenon in question. Therefore, the fact that the tools of the biologist have been applied to analyze a classic physics problem does not break any law or rule. Rather, the endeavor to develop a new qualitative understanding of a frontier of science can be seen as a necessity in achieving true advancement. One of the greatest problems with science education, scientific research, and science as a whole is that people tend to compartmentalize disciplines into mutually exclusive bundles. I've heard physicists describe biologists as "bottom-feeders" and I've heard biologists suggest that they have no need for physics. Both attitudes are self-limiting. The fact that this theory draws support from concrete observations in Particle Physics, Biology, Chemistry, Astronomy, Surface Chemistry, Agriculture and Material Sciences is exactly what makes it strong/sound. Besides, the methods used in this analysis are based on those initially employed by Aristotle, and by subsequent classical thinkers. The fact that we are now dealing with modern questions, modern high-tech tools, and larger systems does not imply that the archaic tools of Aristotle and deductive reasoning are invalid.

---

**Support from Islam**

The birthplace of modern science is often debated. Some would say that the first modern scientist was Aristotle given his pioneering work laying the groundwork for induction, deduction, and inference. Other historians trace the birth of modern science to Baghdad and India where the Silk Route, the convergence of culture, and Quranic directive led to the development of a numeric system. Algebra, taxonomy, alchemy, and medicine can trace roots to early Islamic thinkers. It has already been shown how the Dominium model dovetails with the underpinnings of Aristotle; but it also is supported by the classic methodology of Islam. There is an Islamic tenet that states that "The answer to every question can be found within and/or on the Earth." If this tenet has validity, one could argue that the theory containing analogies to earthworms, micelles, proteins, water, foam, springs, nanostructures, bubbles, fossil trends, fish, shampoo, Swiss cheese, baked goods, and the atomic matrix of a common stone is more compelling than those relying on bizarre mathematical models that have no analogies to anything in nature. In Formal Logic, analogies are only bad for analysis when they don't mesh with the phenomena being described. The match here between analogies used and phenomena described is tight. Most compelling about the analogies used in this paper is the fact that all of them come from very basic aspects of nature. The weakest analogy used was that of the sci-fi warrior. Weak because it does not come from nature, being instead the result of human fantasy—I just couldn't think of a naturally occurring analogue to match the phenomenon I was trying to describe. Although I am not presenting a mathematical model to support my theories, I am confident that a mathematical proof will be forthcoming. Although I, myself, do not know where to begin to devise this mathematical construct, I am confident that the correct solution will resemble the equation for a complex crystalline structure. To the detractors who dislike the lack of math in this paper, I'd reply, that the main problem with the current "solutions" alleged by CPT Violation asymmetry is that this set of theories rely way too heavily on math and has no analogues in nature to confirm its plausibility.

**Resistance and the work of others**

The philosopher Thomas Kuhn identified one common aspect of all scientific revolutions that causes similar backlash responses with their introduction. One common denominator is resistance from the scientific establishment. As Kuhn describes,

> The invention of other new theories regularly, and appropriately, evokes the same response from some of the specialists on whose area of special competence they impinge. For these men the new theory implies a change in the rules governing the prior practice of normal science. Inevitably, therefore, it reflects upon much scientific work they have successfully completed.

To those who had mastered the formulae of Ptolemy, Copernicus was a heretic. Similarly, the initial response to the current work from particle physicists I have known to be strong and angry. One individual, whom I respect, and who wants his name nowhere near this work, was especially vehement in his response. All of the various methods that he attempted to find fault with this model have been included in this discussion section. None of these challenges is against an actual premise or conclusion drawn, and as shown, they do not succeed in dislodging any aspect of this model. The one aspect of his response to which I was least prepared was his statement: "You cannot dismiss the work of 100s of people with those few words, or because you "hate" something." Although Kuhn would predict such a response whenever a new theory is introduced, I was not prepared to hear it from a friend. Also, I do not "hate" anything, really. True, I was never satisfied that the CPT Violation held a clean enough explanation to be correct, but I did not "hate" it. It is also true that I do not identify with the elitism and arrogance that I have observed in some of the Nobel laureates and top-physicists that I have met, but I do not "hate" them. Science has always been an impartial endeavor to me. There is a correct explanation and there are incorrect explanations. Emotions such as love and hate do not enter in. For example, I am the author of this work, though I am not in "love" with any aspect of it. If conclusive evidence is later brought forward that overturns one aspect or another—good—that is the way science is supposed to work. However, if fallacious or inconclusive arguments are brought forward to attack this work, I will defend the Dominium model by showing where the faulty logic lies. Also, I will honor my former mentor's wish and I will not mention his name anywhere within this book, nor will I verbally discuss him or his role in teaching me about the frontiers of current understanding.

I strongly believe that the validity of Dominium model does not mean that the work done by the multitude of individuals who worked on incompatible models was in vain. Probing the frontier of understanding is always an important endeavor. For example, work done at CERN provided the underpinnings of the MRI, which currently aids the sick. Although Al Gore might disagree, CERN's high volume of data was the motivation for the development of the world-wide-web. Telecommunications, data storage, microprocessing, and many other fields have also been positively enhanced by current and past work done at that one facility alone. I realize that the Dominium model might seem to invalidate the life's work of others; this saddens me a great deal. I truly believe that the pursuit of science is the single most enriching endeavor of a person's life. Being wrong from time to time is part of an individual's journey. Though I will be relieved to have finished this document, I do not feel that my own personal journey of learning is close to being over.

---

X

The form of this "proof" is in the tradition of Descartes who decided that when analyzing complex, divergent, and seemingly unrelated systems, one must disassemble all of them to find a common strand. Ever since I was young, the question of choosing a religion vexed me: Hindu, Presbyterian, Catholic, Jew, Shinto, Islam...and others. Why does it follow that I should the religion of my father? After all, you don't choose which family you are born into. What if the family I was born into doesn't practice the "correct" religion? Even the fact that I was born into the family of a Presbyterian minister did not answer my query. Some/many religions warn, "You'll burn in hell if you don't join us," what if they're right? But how to chose? So I did my research to find one thing that the religions had in common to find my answer. My deconstructionist approach yielded the following strand apparently common to all religions that I analyzed (as previously shown in Conclusion 45):

· The Catholics say, "God is omniscient, omnipotent, and omnipresent."
· The Protestants say, "The Lord is my shepherd"
· The Jews say, "God is eternal"
· The Hindu say, "Shiva is ever watching."
· The Buddhists say, "The Path is ever before you."
· The Moslems say, "God is the biggest."

The last discussion on Earth that I ever want to have/conduct is over the question of which religion is better. The reason I don't wish to enter into this conversation is because I just don't *know* the correct answer to that question, nor do I believe it is possible to know the answer to this question. Islam works for me, but who am I to say that I am leading a more true life than anyone else? Supporting this assertion are personal observations that some of the people closest to God that I know or have met have not even been members of an organized sect. Conversely, some of the most evil people I have come across use their assumed superior virtuosity and devotedness as cudgels to justify their harmful actions against other human beings. I believe in a final judgment, where God will eventually chose which of us were good and which were not. No matter how one might delude one's self into dismissing the wrongs inflicted upon others, our life history is continuously being recorded and it's made up of the individual choices that each of us make that affect nature and society. At various points in my life, I've made wrong choices. I knew they were wrong because my conscience deep inside warned me, yet for some reason there was a satisfaction in going the wrong way. To counter my missteps, I've endeavored to live a life maximizing what the Indonesians taught me about "pahalah": [no change in spelling for plural] good deeds that draw favor from God and that help to balance out missteps, essentially the literal opposite of a sin. Acts of pahalah include: learning (esp. science), teaching (esp. to children), appreciating nature in all its forms, fixing problems in systems to make life better for those to come, working hard at whatever one attempts, avoiding consumption of pork, balancing an avoidance of harmful gossip with the joy of laughing at the anecdotes of those around me, aiding those in need, welcoming/respecting strangers, righting wrongs, being thankful of the goodness of the world, and gardening, both on the literal level and on the social level.

To those who might think that I've ruined a perfectly good scientific paper by ending on a note of metaphysics (Conclusions 45, 46, part of 43, and now this), I apologize. My apologies are because metaphysics was actually the beginning point of this analysis—not the end. But isn't that always the truth, beginnings are always endings and endings are always beginnings. On a more concrete note, I do not believe it would have been possible for me to formulate and understand the science within this model have I not been a man of Faith. I don't expect people who don't believe in God to understand such a statement, nor can I truly explain further, nonetheless it is the truth.

The mention of metaphysical understanding is actually necessary within this paper because it will be impossible for many to read this paper without reflection on their own personal tenets. This will be especially true for those practiced in a faith and who have felt and/or experienced the certainty of their personal direction and connectedness to God during points in their lives. After all, this has been a analysis of the mechanisms employed during the creation of "Everything" subsequently predict-ting the observational views of the Universe as we see it today. Some may see this as God's purview. However, not including a discussion on God would make this paper incomplete, and therefore unscientific. I see the work presented in this paper as just another piece (maybe a couple pieces) in the giant puzzle presented to us, as sapient beings, to figure out. I do not think that puzzle (scientific understanding) to be anywhere near completion. I worry about the dangerous potential of Biology, yet I am hopeful that biologists retain their compassion, prudence, and humbleness, inherent in the roots of the tradition of medicine. Though some or many within the modern scientific community may believe there is no longer room for mysticism in the modern study and exploration of pure science, I think Aristotle would have disagreed.

### Alternate site for LHC-type matter-based experiments

The exploration of the unknown is a legitimate use of scientific resources. However, the risks of an experiment need to be taken into consideration as an integral part of its planning. CERN officials might counter that risk assessment has been conducted for the LHC project; it was concluded that any black-hole material produced would be unstable and quickly dissipate in the form of "Hawking Radiation." Any assessment at this time is just another hypothesis (possibly fictional) that is not backed up by proof. Therefore, such assurances are weak, at best.

It has already been put forward by this paper that the potential risks to the planet are too great for the LHC project to continue in its present form—a synthetic stable black-hole will produce the same doomsday sequence no matter where on Earth the initial birth is performed—the Dominium model asserts. Because of the disagreement between theories, it would be extremely unwise to proceed with LHC before a higher level of certainty can be established that categorically determines which of the hypotheses is closer to the truth. In the mean time, Conclusion 44, has already presented a route, which can be taken that will utilize LHC, yet not pose a threat to life on Earth. The creation of antimatter MBH would support the foundations of the Dominium model. Alternatively, if all samples produced quickly dissipate via "Hawking Radiation," then the other hypothesis is supported.

Before more conclusive understanding of the risks involved with matter-based black-hole material can be inferred from experiments that use antimatter, predicting the outcome of LHC experiments is a gambler's bet, not a scientific assessment. If matter-based experiments *must* be done before understanding is more certain, then, for the sake of safety, such tests should be conducted no closer than Venus or Mars. If one of those planets turned into a small black-hole by a failed experiment, it would not exterminate all known life-forms. Actually, the Earth wouldn't be affected at all because a Mars-black-hole would have the same gravitational interactions with the Earth as the normal one does—we just wouldn't be able to see it in the sky any more. Even the Moon should not be risked to such experiments because that object offers mankind much hope for expansion and betterment. This paragraph is the only portion of this work that approaches the classification of "science-fiction." However, in terms of risk assessment, to continue the LHC project on Earth is extremely reckless.

Some may say that Einstein and members of the Manhattan Project fretted about possible catastrophic consequences resulting from the first atomic blast, but the worst didn't happen; one result is nuclear energy, which benefits all mankind. Though such an argument is historically true, it is taken out of context...there was a war going on; it was a calculated risk for the benefit of humanity. Taking such risks now, seems only to serve the betterment of the scientist's ego—with little concern of the potential risk to the entire planet. Again, I seem to be calling in question the work, livelihood, and future of others. This is true, and I'm sorry for any temporary discomfort caused by this work. Yet is undeniable that across the globe there are people doing things, which don't serve the long-term interests of others, nature, society, or themselves. Why should scientists be any different?

**The process of scientific revolution, Kuhn:**

According to Kuhn, "scientific revolutions are taken to be those non-cumulative developmental episodes in which an older paradigm is replaced in whole or in part by an incompatible new one." Kuhn also goes on to identify characteristics inherent in all scientific revolutions. One characteristic that the current work has already come across is resistance:

> For the far smaller professional group affected by them, Maxwell's equations were as revolutionary as Einstein's, and they were resisted accordingly. The invention of other new theories regularly, and appropriately, evokes the same response from some of the specialists on whose area of special competence they impinge. For these men the new theory implies a change in the rules governing the prior practice of normal science. Inevitably, therefore, it reflects upon much scientific work they have successfully completed.

This work is going to meet with strong levels of resistance because it is incompatible with several theories currently accepted to describe naturally observed and experimental phenomena. During the period of Ptolemy, it was believed that all celestial bodies circled the Earth; scientists had devised complex mathematical models to account for anomalies in observations (especially the orbits of the planets) versus the predictions of their theory. Although the Sun-centered model seems obvious today, it originally met with hostility, derision, and resistance when Copernicus put it forward. A similar scenario will ensue once the Dominium model is released. Current theories possess a patchwork of complex math to describe anomalies they face. These models "seem" to explain one aspect of the observed world, yet they do not mesh together well. In the same way the introduction of a heliocentric model overturned prior scientific understanding by offering a better way to describe observations, so too the preceding model better expresses and interprets currently existing data and observations. Unlike other current models, the Dominium model is seamless from end to end. Explanations for all seemingly unrelated tangible phenomena (Big Bang, nebulae, black-holes, solar wind, "dark-matter") have all been explained in a cohesive fashion by the Dominium model.

The preceding work is a new model—the Dominium model. The word "dominium" is akin to the current understanding of a "domain," but not exactly, as was shown. The Dominium model shows a new understanding of the fundamental questions concerning the Big Bang and other phenomena. Current models are at a loss to explain many experimental results and observations of natural events concerning the creation, current condition, and future of the Universe. There are no models currently in existence that can account for observations such as "dark matter," the driving force of the solar wind, the giant black-hole at the center of our galaxy, or the uniformity of the matrix of the visible Universe. The Dominium model describes, predicts, and explains these observations, and many others. The methodology is based on classic Aristotelian syllogisms, rather than relativistic math. Although ancient methods are not quite as in vogue as fancy math, these methods eventually succeed in doing what no other model has ever done: span the distance from Big Bang to Big Bang, including everything that occurs in between. The Dominium model also succeeds in showing the deep supersymmetric relationship between gravitational and electric forces.

---

**Quick explanation of formal logic: functions and limitations**

Unlike past attempts to traverse this frontier, the Dominium model takes a qualitative approach, rather than a quantitative one. Qualitative study was once the classic route for the fields of taxonomy, fundamental algebra, medicine, and philosophy. The methods taken in this construct are the well-hewn ones of Aristotle. This great philosopher was the master of the syllogism. The syllogism is the most formal and most indisputable of all argumentative constructs. It begins with the identification of "premises" which are indisputable and categorical in nature. Words defining categorical statements include "No," "Never," "Always," and "All." Once a categorical "premise" has been identified, it can be used as a cornerstone of any future argument. Two "premises" can be linked together and a categorical "conclusion" can be made. According to Aristotle, if the premises used are 100% accurate, then the conclusion will be 100% accurate. For example:

> All cars have headlights;
> And all headlights can emit light;
> Therefore, all cars can emit light.

Any subsequent argument can use the "conclusion" of a sound syllogism as a "premise" to build from. Usually, this type of continuation of a previous argument takes an "If/Then" format. For example, the "conclusion" from the previous syllogism can be used as a "premise," as follows:

> If all cars can emit light;
> And if all cars can enter parking garages;
> Then all cars can emit light inside parking garages.

All verified conclusions can become premises for future arguments. As long as the premises remain supported, the argument is free to expand. To unhinge a syllogism, all one needs to do is find fault with one premise. For example, the fact that the headlights of matchbox cars do not emit light proves the premise "All headlights can emit light" false. The fault of that single premise will nullify everything that follows. Therefore, close attention needs to be paid to the premises upon which any argument rests.

Often the categorical nature of a thought cannot be concretely established, or as in the case above, outliers may exist. In such cases, a syllogism can proceed on qualitative grounds. For example:

> If all cars have headlights,
> And if most headlights can emit light,
> Therefore, most cars can emit light.
> All cars can enter parking garages;
> Therefore, most cars can emit light in parking garages.

Once the arguments are no longer purely categorical, they become less able to support future arguments. However, that is not to say that generalizations do not have their place in formal logic. Statements where one can state "Most," "Many," "Rarely," "Almost," or which contain a measured sampling percent can be important tools for future analyses. However, when using noncategorical premises, the conclusions reached are no longer categorical. Also, the validity of such a conclusion is always suspect. As shown above, the conclusions reached can also be entirely sound—as long as the number of noncategorical statements used is limited (preferably no more than one). Using multiple noncategorical statements can lead to faulty conclusions. In the fashioning of this model/construct, the conclusions drawn from generality-type premises were tested for their plausibility by using examples from nature. Because this is an exploration of natural events, it follows that analogues to all events described should be found somewhere else in nature. The more universal/common/widespread the analogue is within nature the stronger its case for plausibility.

Sometimes, two different constructs can be used to describe the same question. If the logic used by both constructs is sound, then both will arrive at the same conclusions. However, events do occur where two different individual arguments possess completely different answers to the same question. In such cases, either one or both of the arguments must be invalid. Therefore an analysis of the premises used is essential to decide which construct comes closer to the truth. If any premise is found false, then all subsequent arguments made from that point in the construct must also be discarded. Thus, finding the faulty premise within an adversary's argument is a way to knock-out the competition. This is as true in formal linguistic logic as it is with mathematical constructs. The analysis used in the Dominium model begins with the assumption that its rival, the mathematical construct hypothesis of CPT Violation, has no bearing to its own construction. If both models are "correct," then both will lead to the same conclusions. However, the conclusions of the two models are not similar. Eventually, the faulty premise was tentatively identified within Einstein's basic assumptions with regard to entropy. Because Einstein's theories form the basis of many modern models, including CPT Violation, these subsequent models should all be regarded as false.

Many scientific arguments can be based on only one premise: either experimental or observational data. Although such arguments are strong, they are not impervious, nor are they always 100% accurate. Therefore, experimental data and sampling must be considered to be noncategorical in nature—though some experimental findings are *assumed* to be categorical after repeated tests under varying conditions. Experimental-based arguments can be toppled in the same way as any other, by examining their basic premises. If all experiments and/or observations show one thing to be true, that does not necessarily mean that the conclusions drawn are correct. A classic example is:

> All observed swans are white.
> Therefore, all swans are white.

This argument was found to be categorically incorrect with the discovery of the Australian black swan. The flaw that occurred in the initial argument is that it failed to consider "all" conditions. Sampling is a valuable tool, but if the scope of view is limited then faulty conclusions will most likely result. In scientific research, this is a common cause of mistaken assumptions. Not taking into account "all conditions" is often at the heart of faulty arguments in science. Experimental data always contains and is limited by margins of error, scope of field, and context of its design. Whether those limitations affect the validity of the results varies.

In modern experiments that involve complex machinery, examining the limitations and effect of the machinery on the experiment is crucial in proper analysis. For example, the machine LEP was referenced several times in this work. Knowing that it's an accelerator that shoots electrons and positrons around a giant underground ring is a start. Knowing how the machine achieved this feat is crucial for a proper analysis and ability to make conclusions. Over simplification of the machine's role in this process is a mistake made by detractors in the past in order to attack portions of this model. In such cases, it was shown where the oversimplification was made, by describing the realities of the dynamics of the machine within the situation.

One type of syllogism, which is absolutely not allowed, is one that begins with two noncategorical premises and attempts to draw a general conclusion. Such a fallacious construct can lead to seemingly plausible answers because of the design of the argument, yet be completely false. A classic example of this fallacy:

> Some scientists are egotists.
> Margot is a scientist.
> Therefore, Margot is an egotist.

Examples of fallacies like this are blatant in their flaws. However, similar constructs can be fashioned, be based on two flawed non-categorical premises, and yet appear to fit together, as if were sound. One of the goals of these writings has been to show that this is the mistake of the CPT Violation theory. The CPT

Violation theory can be boiled down to one simple syllogism based on two sets of facts:

a. CPT Violation hypothesis is based on some of Einstein's theories.
b. Some experimental observations seem to show an asymmetry in the decay of B-mesons.
c. Therefore, CPT Violation is proven by Einstein's equations and experimental results.

Both of the premises used in this syllogism are in fact noncategorical. Premise "a" refers to the links between the CPT Violation and relativistic equations, not to the validity of those equations. Normally, this is fine, but if a weakness occurs in the original Einstein equations, then that weakness will be carried up and through all work using those equations as premises. In this paper it was shown that Einstein's assumptions concerning entropy were flawed because they did not take into account the phenomenon of self-assembly. Premise "b" is flawed because of the lack of full scope of experimental trial and because of the lack of predictability of this phenomenon. As it was shown in the discussions of Premise 6, such laboratory "results" are not as conclusive as they are often presented.

Throughout the course of this book, the arguments of detractors have been considered, and shown to be incorrect, either because a premise they used was flawed, their scope was too limited, or an oversimplification or misassumption was made in the test design.

## Appendix 5: Final conclusions

The preceding work is unprecedented. It was written over the two months of summer vacation 2007. I am just an ordinary person, who earns a modest living teaching high school science. There's nothing outwardly special about me: I look normal; I do my job; and I try to act normal. I've taught thousands of children, by now, the basics and fundamentals of the different sciences. I also tried to model for them correct ways to act in a crisis, how to handle an argument, and how to treat each other fairly. I am also a garden hobbyist, especially fond of vegetables, alpines, and annuals. I am not a pedigreed expert on many of the subjects discussed in these pages. Rather I am a well-trained learned mongrel, who knows the subjects, did the research, and presents conclusions. There will be kennel-purists who object to a person of my "classification" being considered at all. Their eyes will be closed to the truth and beauty conveyed in this work. This work is for those who wish to see and consider new ways to interpret current data being generated around us. There are reasons for hope and joy and there are pressing conidérations of grave concern. These are just words, but they are words of advocacy.

The first part of this work is written as a scientific paper. Current journal references are cited throughout where necessary (common knowledge science was not cited). The scientific paper presents a brand new model that outlines and predicts the evolution of the Universe from Big Bang to Big Bang. Along the way, it proves supersymmetry between electric and gravitational forces. Each discussion section contains a formal syllogism to solidify the soundness of each individual conclusion. Discussion follows. For each predicted phenomenon, plausibility was derived through finding analogous situations in nature. If the analogue happened to be from an electrical system, then it was considered to be further proof of supersymmetry.

There is also a second book that was written this summer. It contains messages of hope and methods of living past a "doomsday averted" scenario. Both are published under pseudonym. There are two reasons for this. First, I was to avoid retailation and attention as this scandal began to break. That is the practical reason. The other reason has no practical basis (and actually came first.) I was "inspired" to use the name "Hasanuddin." Actually, that is not truly a pseudonym; it is my religious name that was given to me 26 years ago when I converted to Islam. Until this point in my life, I had never before told this name to any friend, associate, or acquaintance. It no-one's business. Some Moslems may think me ashamed of my religion to continue using my given name long after my conversion. But that would not be true. I kept my given name for social purposes to solve a paradox. Islam teaches to honor ones parents first and foremost. Those two people gave me a Christian name as the first act upon seeing me. To abandon that name would be a grave act of disrespect and hurt against the individuals who I should never disrespect. However, when it came down to authorship of this book, that is not the same as social interaction—it is a different question entirely. Also, I cannot take pure credit for this work, so a pseudonym is even more appropriate. They were my hands that typed; it was me who went to the libraries and combed through current research to back up suppositions; and they are my life experiences; but I still cannot take full credit for this work. Writing this was a non-stop process. I lost over 15 pounds over the last two months just because I'd forget to eat. Revelations and/or images flashed at me, and I typed them down. The faster I typed, the faster the images came. Along with and part of the urgency I felt was the creeping sense of doom. I did not realize the nature of this threat until the typing was right up on top of it. (A very scary thing to experience, let me admit.) I don't know why this stuff came through me, but it did, and that fact can't be changed. I want my life back, but before that can happen I need to make sure that there's an Earth to live it on. This work is hopefully that thing which can assure that possibility.

The onus is now on the reader. Do you agree with the logic that leads to a man-made destruction of the Earth? Or, rather, can you find any fault in the logic that leads to a synthetic doomsday? This question is crucial. The fact is, this question hangs in the balance within the near future. If we are heading toward our own total destruction, it makes sense that this information should be passing through me at this moment in history. Think about it:

a. If God truly loved the world and all its creations;
b. And if all creation was in absolute danger of extinction;
c. Then, now would be a time that God might intervene.

It also seems logical that as part of that intervention, the answers to some of the greatest science questions of all time and a handbook on how to live long into the doomsday-averted future would be included. The science revelations are a sweet-tener for the bitter medicine-news that the world's biggest cooperative project needs to be stopped and/or diverted. The handbook, the Dominium Epiphanies, will be a gift for all who live from now and into the future. Together these things are presented for the reader to digest/apply toward future junctures as they see fit.

In the preceding pages a complete model has been delivered. All loose-ends are tied up and what was once unexplainable is now given reason: dark matter, black-holes, antimatter, solar wind, nebulae, etc. Science, and scientific exploration, now stands at a precipice. If we continue (experiments at CERN, Brookhaven, and other labs) then there is the very real danger of losing the entire planet. Can we abstain, stop, and alter our course? Or, are we mindless lemmings chasing paths without foresight or consideration? The answers to these questions will be answered one way or another in the very near future.

Disks containing this paper were sent out to world leaders across the globe. Did they understand the danger? Did they even open the disk? As a reader of this work, you might be able to tip the balance. Contact your governor, legislator, union chief, president, and whatever influential individual you might know, tell them of the existence of this book, and have them act to protect us all. Also, talk to your friends, family, and all the other un-influential people you know and encourage them to read this book for themselves. The future is not certain. CERN is just a laboratory run by humans. If enough pressure is generated, stable black-hole material need never be created on this planet.

Concurrent to the transcription of this model, I, the author, underwent a profound religious reawakening. Two books were actually written this summer. The second book, the Dominium Epiphanies (forthcoming), contains my own understanding of metaphysics and my own insignificant place within this world. I am much less certain of the categorical nature of what is transcribed in that section, because it cannot all be verified through syllogisms and formal logic. However, the same feeling of clarity that guided the writing of the Science Section was present throughout the writing of the Epiphanies Section. Because the Epiphanies section will be of value only if doomsday is averted, will not be published unless LHC is stopped.

There will be those who resist these theories because they are blinded by their own egos, prior misconceptions, or refusal to read all that has been transcribed. Such people will never be convinced, and they may stand in the way of those who do see the truth and danger. Their blockades must be sidestepped and overcome. You cannot please everyone, especially if you are trying to do what you know in your heart is correct.

If the experiments at CERN are not halted, and if stable black-hole material is created on Earth, then the cascade of subsequent events will be unstoppable. As individuals, we are insignificant and seemingly powerless. At times over the past couple months, the burden of what has been revealed almost made me give up. Yet, unexplainable coincidences have put me back on course. Most recently, my childhood neighborhood was in the national spotlight, in the "La Jolla landslide." I am not sure if the actual home I grew up in was one of those destroyed, but the main street destroyed was the one I grew-up on. The irony of this "disaster" put me back on course. The destruction on Soledad Mountain Road is nothing compared to what will happen if CERN scientists are successful in creating stable black-hole material. At worst, my old neighborhood has been altered. Compared to the idea of being compacted out of existence, it is completely insignificant. There is still time to alter the course we are on…but time is running out. Although death is a certainty for us all, the death of the Earth need not happen within our lifetime.

The maiden start-up for LHC was set for Sept 2008[6]
There is <u>very little time</u> to act.
*This date has moved at least seven times since the writing began for this book. If the day you read this is past the date above, that does not necessarily mean we are out of danger. As long as they are rebuilding and preparing, we are still at risk.

---

[6] See the Press Release on the next page to see what appears to have happened

FYI: The following Press Release was sent out onto **AP, AFP**, and other wire services on Sept 24h, 2008. Today is the 27th. What this **release** might trigger, if anything is yet to be seen.

===============PRESS RELEASE===================

Forget the lies of our trusted politicians or of the security of banking institutions, examine what our esteemed scientists have been weaving. Amid fanfare of the $6,600,000,000 LHC-machine's "successful" start-up on Sept 10th, within a few short hours later it broke. Why wasn't this reported to the public on the day it occurred? For over a week their silence passively and aggressively allowed the world to believe that LHC was running smoothly when it was actually down and broken. The duplicity of not immediately owning up to such errors weakens the legitimacy of the entire project.

When the machine broke, the CERN facility was crawling with reporters, observers, and TV crews. Yet not one word was mentioned that the colossus was not functioning? Ironically, on Sept 12th (a day after the multi-level failure) a story was floated across the internet about a group of Greek hackers that allegedly infiltrated the CERN computer system. Of the CERN officials and scientists interviewed for this story, one admitted that it was a "scary experience." None of those commenting would admit that there was a much bigger story unfolding—LHC was inoperable and offline due to systemic failure, not hacking. LHC "quenched" as scientists would later describe, as if it were just an expected glitch in their otherwise perfect masterpiece. But whether you use a fancy word or not, busted is busted. How can anyone believe that it is normal or expected for a machine to seize up in the first minutes of operation and spew out cryogenic material onto the lab floor resulting in hundreds of instruments and circuits to heat up over 100 degrees within moments? And how could the scientists interviewed for the hacker story forget to mention the real story of their broken machine?

A similar pack-mentality CERN reversal can be seen with the whole question of LHC's ability to produce dangerous black-holes. In 2000 I was sent by NSF to represent the USA in CERN's annual HST program. As part of the program I attended dozens of lectures. One lecture concerned the "wondrous possibility" that LHC might generate man's first synthetic black-hole, thereby guaranteeing plenty of future Nobels to European scientists. When the obvious question came up concerning risk, it was assured that only massive black-holes could be stable because the only "known" black-holes were the result of the collapse of massive stars. Today, members of the scientific community are coming forward with models, concerns, and arguments that suggest that even mini black-holes could be stable (and catastrophic.) In response CERN has changed its tune. To hear them talk today is a complete flip-flop from what was being said in 2000—now they say that it is impossible for LHC to be unsafe in any way. One moment they boast of a future achievement, the next moment they suggest it as an impossibility. Did the Science change, or was the only calculation to change one to soothe the public?

---------------------Links to author background and facts cited[7]------------------------
First hand experience at CERN –
http://teachers.web.cern.ch/teachers/hst/2000/people/paul.htm
Extent of the actual damage to the LHC machine –
http://news.bbc.co.uk/1/hi/sci/tech/7626944.stm
http://www.sciam.com/blog/60-second-science/post.cfm?id=lhc-helium-leak-will-shut-collider-2008-09-20
Duplicitous release to AFP one day before admitting the machine was down for two months – http://www.news24.com/News24/Technology/News/0,,2-13-1443_2396209,00.html
Hacker story coverage during the silence of LHC failure –
http://www.engadget.com/2008/09/12/hackers-hit-lhc-computer-system-deemed-scary-experience/
http://www.mxlogic.com/securitynews/network-security/lhc-hack-highlights-network-security-risks925.cfm

I, Paul Eaton, used the pseudonym "Hasanuddin" (to avoid preliminary attention and/or retaliation as this scandal broke) to author *The Dominium* a scientific model that shows mini black-hole stability –
http://www.webster.it/book_usa-dominium_sequencing_antimatter_and_gravity-9780980096330.htm
http://www.seekbooks.com.au/book/The-Dominium-Sequencing-Antimatter-and-Gravity-Effects/isbn/9780980096330.htm
http://www.myshelf.com/miscellaneous/08/dominiumsequencingantimatterandgravityeffects.htm

PS. Please use this information/story/letter/foundation however you see fit.

---

7 As of today, September 27, 2008, all of these links are operational. However, as was the case with the SciAm blog, it would be unwise to believe that will always be the case.

**Last: Revelations**

I do not like that word, nor do I like what it implies. However, coincidences of the last three months are too weird to use any other. I wish none of this were happening, but it is. I wish to be left alone and to live my life as just another ordinary high school teacher, but I fear that won't happen. There is urgency in matters at hand.

Coincidence, circumstance, and happenstance put me at three in-depth trainings: CERN, NASA, and MIT. I learned from the most knowledge-able scientists available about the frontiers of modern science. During the most recent training, at MIT, I was "hit" with something unexplainable and these words have been pouring out ever since. When I started typing, I had no idea where the words were leading. I knew all along that this was urgent. I knew that the future held two very distinct possibilities: hope, enlightenment, and cooperation between people of all nations or doom. I had no idea of the source of the danger until the narrative reached that point. The danger posed by the manmade creation, LHC, is real and plausible. If that danger is unleashed, there will be no future generations and all Earth's life-forms will become extinct.

When the scientific truths were first being passed through me I was receiving alarming signs in the form of numbers. Cash register receipts and dockets were all lining up: $18.18, $9.87, $22.22, $3.57, $5.67, $20.00, $44.44, etc. It was scary going to the store, because every day, no matter what I bought the receipts contained numbers that didn't seem to appear by chance.

The number of ironies and coincidences that surround this work or have been made apparent over the last six months is quite amazing. True, individually each of these little observations appear to be trivial, but when does the number of mounting ironic coincidences cease to be inconsequential? For example, where do I live: in a town, also known as, "Dot" in a state, also known as, "Mass." This never occurred to me until I was preparing to insert my return address for the mailing to foreign leaders, however, at that moment the irony hit me, as did a call to put "Dot, Mass" as my return address. Somewhere along this journey I stopped discounting ironies and coincidences. After my submission, if I felt a certain push to move in a particular direction, that's the way I forged forward.

The call to write was pushed upon me while I was still participating in the MIT lecture series. Not wanting to give up the chance to learn more science, I attempted to do both activities, which meant that I had to ride the subway often. Another uncanny coincidence: every time I descended to the subway platform, an Ashmont train was either waiting or just arriving to the station (unheard of normally). That occurred a remarkable ten or twelve times in a row. These, and other very strange coincidences, convinced me of the urgency and I wrote and wrote as fast as I could. The more I wrote, the more was revealed. My friends described me as a man possessed. I've lost about fifteen pounds as a result of basic lack of hunger. This work took precedent over all else. Once the foreshadowing of the LHC-doomsday was conveyed, my writing continued. Surprisingly, I am not yet tired. Driving me is a concern for all the students I've taught and for the trees I've planted. I guess those are the things representing what I had always believed to be my main imprint on the future of the planet.

When does the number of coincidences cease to be coincidental? This is a truly important question! I am naturally a cynic and a skeptic, but I have been overwhelmed to submission by the ceaseless flow of irony, coincidence, and sign.

The most ominous implication of this book is the possible apocalyptic sequence caused by a stable black-hole. Because the major topic of that implication is apocalypse, the book of Revelations comes to mind. My father, a Presbyterian minister, never preached from that book—he didn't like it. He once told me that it didn't go with the rest of the Bible and there was nothing good that you could tell a congregation from it. Maybe. Yet, at a time like today, it needs to be referenced.

Probably the most famous quote from Revelations has to do with 666. Hollywood has made a number of movies concerning this number, and an even greater number of books have been written. To those who get their religious information from the cinema, they believe that 666 is the sign of the devil/Satan—that is how most of Hollywood portrays it. However, that is not what the Bible says. The Bible simply states in Revelations 15:18 that 666 is, "the number of the Beast, for it is the number of man." Other translations have it as "the Mark of the Beast." But, what is a "mark"—a moniker, logo, diagram, emblem, chevron, sign, indication... Okay, the most ominous implication of this book is that the stable black-hole (Beast) that could easily cause a Revelations-like fire & brimstone End might soon be made at CERN. View the design of CERN's logo for yourself. How many "true" sixes do you see? Personally, I see three over-lapping sixes and two things that aren't sixes. Does this mean that CERN is the home of the devil?—absolutely not, nor does it mean that there is anything evil about that facility. Hollywood is wrong. Look back at your Bible; you will not find another place where the "Beast" is referenced. No, the Beast only has to do with the possible end of the world; Satan is some-thing altogether different.[8] To have 666 show up in the logo of the facility implicated as the potential birthplace of the Beast is almost expected.

The Four Horsemen of the Apocalypse is, yet another, famous reference to the book of Revelations. Indeed, there was a huge coincidence here, as well. The Scientific American community blog-space has been the battle-ground for much of my efforts. I am so thankful to have found that site. If my efforts are successful, much credit is owed to Scientific American. A couple days after starting my blog, I was touring through the sites of others to see if there were any people writing

---

[8] A topic that will be explored in-depth within the Dominium Epiphanies, to be published, if and only if, LHC is stopped.

about similar topics who I could invite over to join the discussion that I was starting. When I reached the Quantum Poetics blog, the most recent entry was:

> <u>The Four Horsemen of Physics</u>
> Horsepower is The Four Horsemen that are The Four Forces that will come together at the end of the world: conquest (gravity) rides a white horse; war (electromagnetic) a red horse; famine (strong force) a black horse; plague (weak force) a pale horse.

Honestly, my blood ran cold as soon as I laid eyes on it. Aside from the first word, "Horsepower," everything else lined up. Replace the word "horsepower" with "Black-hole formation" and it reads perfectly in line with the predictions of this book. True, there are many theories concerning black-hole formation, but I imagine that the stability of the black-hole is achieved through the "coming together" of the four known physical forces of nature. In concert, and in close proximity, maximum irreversible stability is achieved. According to this model, if CERN creates a stable sample, the End of the world has been predicted.

Another Revelations reference is the war between Good and Evil. Perhaps being fought at this very moment. "War" need not mean an armed bloody conquest. In the attributions section, I recognized Dr R. Philip Eaton, for his encouragement and belief in my abilities and model. His very first words to me, with regards to advancing new theory upon the scientific community were, "Be prepared for war, because that's what it is." And so it has become. The cyber-attack on Scientific American and the subsequent multiple hacking of Hasanuddin.org are nothing short of war-moves. On their side are proud scientists and bureaucrats unwilling to admit that they could have ever been wrong. They are motivated by greed (their salaries are quite large), fame (they envision Nobel gold), sloth (they do not want to consider new options), and egotistical pride. On my side is an army of children. These volunteers have helped write letters, distribute the electronic version, create the Hasanuddin websites, and comb the web for allies and defenses. They are driven by a love for life and a deep sense of compassion for others. One girl got very emotional while writing to the Prime Minister of Chile; she asked me how many people were living on Earth. As the number began to sink in, she wept. Not for fear of herself, but for all those other children around the world who are in danger as well.

Another coincidence from Revelations is: men who "calculate" the number of the Beast. This passage need not mean that the answer to the calculation is actually 666. As one evangelical friend has told me that this was always interpreted to mean accountants, bankers, and/or people who put money in front of God. But also this may be reference to the mathematical modelers whose calculations justify the LHC machine. All of the safety assurance arguments are based of theoretic computer simulations, which are based on the assumptions and calculations of men. From these calculations the LHC machine was fashioned and its safety justified. If, however, those calculations are flawed, then there is no justifiable safety.

Let me stress, CERN and its scientists cannot be considered inherently evil; however, the motivations behind the current conflict might be. Also, the children who are fighting on my side cannot be considered inherently good; they are just as prone to temptation as anyone else, but their motivations to volunteer to assist in the current campaign, do appear to be whole-some. I firmly believe that no-one can live a perfectly good or evil life, yet we are all as worthy as any other.

As far as the logo is concerned, I mean no disrespect—it is quite handsome and clever. The "6" design represents the heart of the type of accelerator used at CERN. The stem of the "6" is the linear accelerator that injects accelerated particles into the loop accelerator. Yet, it is an extreme coincidence that designers chose to use three overlapping 6's. Before this summer, I never noticed it. But now, it's hard not to see. Again, that does not mean there is anything "evil" going on there, yet if this machine truly could launch an End-of-the-world sequence, then Biblical omens might be expected.

Will Good triumph over Evil? Will LHC be stopped? Or will Power and Money triumph over Hope and Innocence? The future is not certain. It could easily go either way. Personally, I will not let apathy, fear of retribution, or insecurity of self let me slow down until the outcome is certain. For me there are three possible outcomes:

1: An Achilles' Heel can be found with my scientific reasoning. This out-come would be "embarrassing," perhaps; however, the worst that it would mean is that my life hasn't changed, tomorrow will always come, and we can all go on as before. Yet, the longer this model is circulated, the hope for this option diminishes. Hundreds of the first edition have already been sold. The abridged electronic version has been freely given to thousands of people, either directly or as a free download. True, there have been those who have attempted to find fault with the model via the Scientific American blog-space, yet all those who have posted so far hadn't really read the model and their attacks were based on projections and false assumptions. None of the attacks validly addressed a single issue within the model. With each passing day, hope for this option diminishes.

2: LHC is stopped. Remember what I wrote in the beginning of this book: at the time of writing this model, I felt two welling emotions. Both emotions are totally illogical and had nothing to do with the science being transcribed. However, these emotions do pertain to the crossroads at which we find ourselves. The first emotion pertains to this option—overwhelming joy, happiness, and reconciliation between all the tribes of the Earth. This could be the dawning of a new age of prosperity and mutual under-standing and peace for all humankind on all corners of the globe. Science will catapult forward; society will reach levels never before attained and goodness will abound; and a time for us to rejoice.

3: LHC creates the Beast. This option would justify the second welling emotion that I felt while initially transcribing this work: doom, dread, des-pair, and death. In addition to the earthquakes, volcanoes, and tsunamis that would be caused by the slow implosion of the Earth, societal structures and protections would disintegrate. The whole world would come to know what had happened and what their fate was—anarchy, bedlam, and conflict would ensue. Government, financial institutions, and police will collapse. People will hoard the limited food available; others will starve. Neighbor will kill neighbor. All the while, the frequency and severity of "natural" disasters will increase and extinguish people in-mass.

Some may say that there is a $4^{th}$ option: LHC is a dud. That is wishful thinking. I don't believe it. Seven years ago, I sat in a CERN lecture hall and heard how this powerful new machine was designed. In that lecture, it was stated quite clearly that one of the major hopes for this project was to win the race to create the first synthetic black-hole material. This lecture is 100% opposite the current propaganda that has been issued from CERN. If you believe their current statements, this project is quite small and benign. But how can anyone believe such statements? If it was so small, why did it cost $6,800,000,000? If it is so benign, why will it use more power that the giant city of Geneva? Also seven years ago, I was fortunate enough to go on a tour of the LEP tunnel (the current home of LHC). The technician told us that we were among the last to ever be allowed into the tunnel. "Once LHC is running, so much radiation is going to be produced that this tunnel will be permanently contaminated." What kind of small and benign project produces such high levels of radiation? No, I do not believe the propaganda. If LHC is allowed to run, it will not be a dud.

If the End does occur within my lifetime, I am not fearful—not for myself, at least. Although, I don't relish the pain that might be associated with my own death, I am not afraid of being judged by my life's work. I've made mistakes, gone astray, and upset the balance a few times in my life, however, on the whole I hope to have given more to society than I have ruined.

When I arrived at work to begin the school year, I had another moment of clarity oddly comforting with respect to the End of the Earth. As I mentioned, initially the process of writing was driven by a fear, not for myself, but for the children that I had taught and the trees that I had plant-ed. Although, I have planted many trees in my life, I happened to notice my two most recent plantings (a hawthorn and an heirloom apple), which were planted in the dead of night in the schoolyard as a silent gift to the neighborhood and future students. I planted them this spring and had been tending them before beginning this all-consuming project diverted my attention. I had envisioned the hawthorn, which was planted between the sidewalk and the street, to grow into a beautiful flowering shade tree with colorful berries that brighten the day of pedestrians. The apple was planted in a spot that aligned with a row of preexisting trees. I imagined it would bear fruit that the kids could eat, throw, and enjoy. Upon return from the summer, my heart sank to see the apple: its leaves were dried and curled. The August drought had severely weakened the poor tree, because I forgot to water it. At first glance, the hawthorn seemed green, healthy, and un-affected by the drought. However, closer inspection of its base showed that it had been girdled by careless misuse of a weed-eater. Both trees are poised to die. End of the world or no end of the world, neither tree may live to see the next spring...and that's "okay." When I planted these two, I knew full well that their chance of survival was not great: adolescents are mindlessly destructive, snowplows could uproot them without even knowing they're there, and disease is always possible. Yet I had a dream, hope, and image of what they might have become. That vision was a basis justifying the urgency and need to write. Now that the project was nearly complete, there is a great irony to find these trees poised to die anyway.

To me, this irony is a "sign" with meaning. It's okay if the trees die. It's also "okay" if all of my former and current students die—they were all going to die anyway. As a gardener, you can only do so much. In the same way, it's okay if the planet dies. It was going to die anyway. As the bringer of these messages, I can only do so much. I believe that if these warnings are not heeded, and the course is not corrected, then a synthetic black-hole will be created that will ultimately consume all life on Earth. That would be a very sad turn of events, yet one which will not be lamented. Although people would die in mass, they were all certain of death anyway. Nothing, except for the length of individual lives, would have been changed. After all, it's just the end of the Earth (as we knew it), not the end of the Universe.

Being afraid of death and being worrirfd of the Last Day (Judgment) are two different things.[9]

---

[9] To be discussed in the Dominium Epiphanies

**Appendix 6: Science Abstract**

The following is a logical construct based on one altered assumption, namely, what if "like attracts like and opposites repel" holds for gravity? How would that affect our overall understanding? Using formal logic, fundamental principles from physics, experimental data results, observation records, and analogies from nature (to give plausibility for conclusions) I have transcribed the Dominium model.

Outline of the Dominium model:

1) A mechanism explaining why the Universe expanded extremely rapidly initially and why it's still expanding though at a slower rate. {F proportional to $d^{-7}$ [or even more extreme] changing to F proportional to $d^{-2}$ as distances increased, occurring simultaneously with the Dark Event adiabatic availability of internal space.}
2) A mechanism explaining why embryonic galaxies did not collapse in upon themselves. {Opposite-matter micelle puffing up the forming galaxy}
3) A mechanism positing and explaining the existence of a giant black-hole at the center of our galaxy. {Micelle movement away from that point, but not through that point, leading to collapsible high quantities of mass. Micelle-foam collected and eventually stopped the growth}
4) A mechanism explaining the nature of "dark matter" {MBH traveling between galaxies.}
5) An explanation for why "Dark Matter" does not pass through Earth or has not been detectable on Earth {Opposites repel}
6) A mechanism explaining the production of a diffuse dust cloud for the early remaining matter of the Milky Way {Simultaneous sudden reduction in volume taken up by micelles as they collapse into MBH stretching space-time across the whole of the interior of the matrix.}
7) A mechanism for the apparent flat event horizon {Matter warping space-time oppositely in manner for antimatter versus matter, leading to a net zero effect on overall continuum.}
8) A prediction that matches observations of a uniform and patterned mass distribution of the Universe outside the Milky Way. {Crystalline super-matrix of galaxies.}
9) An assertion for the point at which Einstein's model might be flawed {The process of self-assembly defies current Einsteinian assumptions regarding entropy, i.e., in this process a primary mixture tends to self-organize into a more ordered state, "choosing" to move away from chaos. As a primary mixture of dominia, the galaxy matrix would be expected to undergo self-assembly.}
10) Simplicity, beauty, and elegance {Analogues of all predicted repercussions exist in nature, and, most important, CPT *symmetry* is preserved.}
11) Explanation for the lack of recorded evidence concerning the titer of positrons within the solar wind as measured by NASA's ACE and WIND programs. {Immiscibility is in effect at the point of data collection. Any

attempt to move the collection device into an antimatter stream causes the fluid boundary between matter and antimatter to move accordingly.}

12) An explanation for the sharp boundaries observed in solar wind plasma structures. {The overall structure of the solar wind would be expected to form alternating fronts of matter and antimatter radiating outward. The data-collecting spacecraft acts as a conduit for matter to pass by. Antimatter streams, conversely, will be steered away from the concentrated matter of the space station due to immiscibility. Spikes appear with the arrival of the next front of matter being moved.}
13) A mechanism and prediction for the creation of nebulae. {In a galaxy like the Milky Way, the matter portion of the solar wind, from multiple stars, is unable to continue the journey out of the dominium and settles in areas of gravitational eddies.}
14) A continuation of the consequences of the Dominium construct to describe how the solar wind sets up the conditions between dominia which will result in a continued decrease in the expansion rate of the Universe and ultimately lead to the next Big Bang {Loss of positrons from our galaxy leads to an increasingly negative charge, symmetric process in antimatter-dominia, leading to electrical-attraction forces eventually outweigh gravitational repulsion forces between dominia, leading to a collapse of the entire system.}
15) A proof that the Dominium model shows supersymmetry between gravitational and electric forces. {Self-assembly, Big Bang/black-hole, and anecdotal evidence.}
16) An explanation for the event that flattened a forest in Siberia {Near-Earth encounter with a small MBH, a.k.a., "dark matter", opposites'-repulsion force flattened the forest as the micelle was deflected away.}
17) A call for a moratorium on further studies seeking to create matter-based black-hole material on Earth {Doom}
18) A way to save the LHC Project from the scrap-heap. {Convert LHC from a proton accelerator to an antiproton accelerator, thereby creating antimatter black-hole material rather than matter-based (Earth-consuming) black-holes}
19) Proof through the construct itself of supersymmetry between matter and antimatter. {Ten separate examples of supersymmetric equivalents between known electrical and gravitational systems.}

## Appendix 7: Delineation of the Construct
**(Its Premises, Conclusions, Blogs, and Discussions)**

**This portion of the book lists the premises used and the conclusions and discussions that resulted. It provides a quick and easy reference for the reader to evaluate, consider, and refer back to other sections of the book**

## Glossary of Terms

**Antimatter** – Refers to a genus of particles that are the mirror twin of more commonly known matter. For every particle of matter, there exists a mirror equivalent of antimatter. For example, positrons are the antimatter equivalent of electrons; antiprotons are the antimatter equivalent of protons; antihydrogen is the antimatter equivalent of hydrogen. Hence, for every known particle, theoretically there exists (or can exist) an equivalent structure of antimatter. For two mirror twins, the mass is exactly identical (or roughly identical, depending on which study, theory, or degree of error is used.) Also the magnitude of electrical charge is identical between matter/antimatter pairs, though if one is (+) the other is (-).

The notable exception to this rule are photons, which do not have a mirror twin because they are the antiparticles of themselves. Therefore photons cannot be correctly designated as either matter or antimatter, rather they represent something which is in between/both.

According to the implications of the Dominium model, antimatter is projected to account for 50% of the mass of the entire Universe. (See Premise 9: Parity exists in all forms)

Contrary to popular assumptions, antimatter is actually quite common and a natural member of the near-Earth Universe. Tons of positrons are created in our Sun every second.

**Annihilation** – An event that occurs when matter and antimatter are forced into the same location. In this event, mass is converted into energy as described by the reverse application of Einstein's equation, $mc^2=E$.

**Anti-gravity** – This term is never used within the Dominium model. It is noted in this glossary because some previous theories do use it. However, its use is not standardly applied and possesses inherent linguistic ambiguity. Essentially, this term means "gravitational repulsion" (the phrase that is used by the Dominium model.) However, because the same term has been extensively used within sci-fi, from Buck Rogers to Star Trek, the intended meaning becomes blurred because of this baggage. To avoid confusion, only the phrase "gravitational repulsion" is used to describe the interaction between matter and antimatter.

**Black-hole** – For the basis of the model, the simplest definition of this structure is used. Essentially a black-hole is a configuration of particles where there is little or no space between adjoining particles/members. Contrast a black-hole to a "normal" rock, which is comprised of more than 99.9% void space between particles, black-holes are incredibly dense. The incredible density of black-holes leads to gravitational vectors that are incredibly strong at increasingly close ranges. This exponential increase in gravitational intensity helps explain how/why black-holes are able to consume adjacent material. (See Premise 4)

Another aspect that is both central and notable of black-holes is their incredible stability. This heightened level of stability is to be expected when they are viewed as the fifth and final phase of being: plasma, gas, liquid, solid, and finally black-hole. When going through the sequence of phases of being density increases, kinetic energy of particles decreases, stability increases, space between particles decreases, and stabilizing interaction between neighboring particles increases. Empirical data confirms this placement of black-holes as being the fifth and most stable form of matter in that there is no confirmed observation of any black-hole spontaneously destabilizing into something else. Such empirical data is in direct conflict with the prediction of "Hawking Radiation" (refer to glossary entry for Hawking Radiation for further discussion.) (For further discussion of black-hole stability see Conclusions 18,40,41, and 47)

**CMB (Cosmic Microwave Background)** – This is the "First Light" of the Universe. Once the Universe cooled to around 3000 degrees Kelvin, electrons could begin coupling with protons. Once that occurred, photons were produced as a byproduct. These photons would have been produced throughout the Universe wherever and whenever there were both protons and electrons present and conditions had cooled below 3000 degrees Kelvin. One of the most surprising observations of the study of CMB was the observation of extreme uniformity of the intensity of CMB regardless of where in the sky the measurements were drawn. (See Conclusion 48 and Appendix 2)

**Culvert Black-hole** – See SCGBH

**Dark Event** – One of the most profound and important events in the early devellopment of the Universe. The Dark Event occurred either subsequent to or overlapping the moment of the initial establishment of Immiscibility. During the Dark Event both MBH and SCGBH were formed. The sudden formation of these two very dense types of structures suddenly opened up huge amounts of internal volume for the Big Bang fireball to expand into without drastically increasing overall size. Therefore, the Dark Event can also be thought of both as a time of sudden compression of some material and as a moment of adiabatic expansion of the non-compressed material, i.e., the material remaining that was not compressed into either MBH or SCGBH. Because conditions of the Dark Event were adiabatic (or very close to adiabatic) the net affect on the remaining matrix was sudden cooling.

The combination of sudden availability of internal volume and cooling of the non-compressed material accounts for the inflection point of the accepted Hubble Expansion Curve. Therefore, the Dark Event would have occurred in the moment of time that overlaps with this inflection point. (See conclusion 10)

**Dogpiling** – A theorized interaction within a like-like supercollider, e.g., LHC, where subsequent collisions cross-interact with one another with the possibility of producing an accidental sample of black-hole material.

**Dominium (plural: Dominia)** – The definition of a dominium is not the same as that of a galaxy, though they appear to be similar. This difference is most easily viewed through analogy: atomic boundaries are to nuclear boundaries as dominial boundaries are to galactic boundaries. The radius of an atom is enormous when compared to the radius of its nucleus. So too, the radius of a dominium is similarly enormous when compared to the radius of its galactic cluster. In both cases, one is considering roughly the same particles, however, the field of view is much different. Just as viewing on the atomic level one must consider the dynamics of electrons, so too viewing on the dominial level forces one to consider the dynamics of MBH and positrons carried away on solar winds via gravitational repulsion and gravitational attraction.

The identified constituents of a dominium are: the core supermassive central galactic black-hole (SCGBH), the micelle-foam covering the event horizon of the SCGBH, the main cluster stars and solar systems that rotate around the SCGBH, nebulae repositories of like-type material carried outward by the solar wind, opposite-type MBH traveling outward from the core, opposite-type solar wind particles also traveling outward, and the possibility of like-type MBH and like-type solar-wind particles originating in adjacent dominia having crossed the dominial boundaries and heading towards the SCGBH.

The proto-galaxies, or dominia, that were initially formed immediately after the establishment of Immiscibility between matter and antimatter and the Dark Event (production of MBH and supermassive central galactic black-holes.) The boundaries of a dominium extend far out into space away from the galactic cluster. Boundaries are set where the gravitational influence of one galactic cluster outweighs the influence of its neighbor. At the boundary, influence would be split 50:50 between adjacent dominia.

**Gravitational likes** – Two particles; either both are composed of some form of matter, or both are composed of some form of antimatter. In proximity to one another they will experience gravitational attraction.

**Gravitational opposites** – Two particles; one is composed of a type of matter and one is composed of a type of antimatter. In proximity to one another they will experience gravitational repulsion.

**Gravitational repulsion** – The interaction between gravitational opposites (antimatter vs. matter) as set forward by the central hypothesis of the Dominium model.

Haw**king Radiation** – An unverified theory suggesting that mini black-holes might spontaneously and harmlessly evaporate away. However, there is no observed data of this process ever occurring. The notion of Hawking Radiation is the primary justification of the safety of exploratory projects, like LHC. Ironically, proponents of Hawking Radiation hope that LHC might supply the first evidential proof to support this theory. Hence forming a mutualistic and circular relationship between this theory and projects such as LHC. If the theory is flawed, then LHC could easily be unsafe... but we won't know for sure until LHC is at full power. (See further discussion on page 131.)

**Immiscibility** – 1: Permanent separation between gravitational opposites; 2: A moment in the development of the early Universe when direct interaction between matter and antimatter was no longer possible due to primary gravitational self-assembly; 3. A state of permanent separation that is achieved once any extremely heterogeneous system self-sorts in response to decreased energy levels. For example, immiscibility is reached between MPP and the rest of the solar matrix as MPP travel outward away from the solar core into regions of decreased particle density and overall energy. Once immiscibility has been established, it becomes impossible for a naturally occurring annihilation event to occur. (See Conclusion 2)

**Imperfections** – A term originally coined by Guth to refer to the idea that the initial Big Bang fireball would not be 100% evenly distributed and that there would be a random variability to the distribution of material. Ultimately, according to Guth's theory, these imperfections led to the creation of the galaxies. The Dominium model takes Guth's idea a step further. The imperfections would either be areas that by chance contained a larger proportion of either matter or antimatter. These imperfections led to the creation of the dominia, which subsequently went through the process of self-assembly, and then ultimately evolved into galaxies. (See Premise 7.)

**LEP (Lepton Electron-Positron collider)** – A particle accelerator housed underground at CERN that used one beampipe to accelerate antiparallel traveling packets of electrons and positrons into each other. LEP was in operation until 2000 when it was dismantled so that the tunnel in which it was housed could be used for LHC.

*According to the Dominium model operation of LEP was "safe" because it accellerated packets of matter particles into packets of antimatter particles. Because of this dogpiling could not occur, therefore there was no real risk of ever accidentally generating a sample of black-hole material. The most common type of interaction was therefore annihilation, which is a relatively harmless, and now well understood, release of energy.

**LHC (Large Hadron Collider)** – A particle accelerator housed underground at CERN that uses two beampipes that strategically cross paths at engineered junctions to force antiparallel traveling packets of protons (or stripped heavy ions like Pb or Au) into one another. The start-up date of LHC has changed many times as engineering flaws, design flaws, and other quirks have hampered its progress. At the time of writing this glossary, the machine was almost two years behind schedule. (See pg 155)

*According to the Dominium model operation at LHC could be extremely unsafe. Because that machine will be slamming successive packets containing upwards of

10,000,000,000,000 protons (or heavy ions) into one another there is the possibility of cross-interaction, i.e., snowballing/dogpiling, of particles in one place. If this does occur there is the possibility of the formation of a mini black-hole. Since black-holes have been identified as the most stable phase of being and since mini black-holes are expected to exist in the form of MBH, therefore a synthetic black-hole would be predicted to be stable. However, there is also the possibility that the LHC machine will not be strong enough to produce any black-hole material at all. In such a case, LHC might be able to operate safely, but a next generation more powerful machine would not. In any event, the exact degree of safety of the LHC machine cannot be precisely determined before it is actually set into operation.

**MBH (micellular black-hole)** – A type of black-hole that was formed early on within the development of the Universe. It is expected that MBH were formed within each proto-galaxy (dominium) at the same time (the Dark Event.) No matter where they were formed, the MBH would be of the opposite genus of material as the parent proto-galaxy. For example, in the forming Milky Way, the MBH formed would be composed of antimatter, while in an antimatter-based proto-galaxy the MBH formed would be made of matter.

Because the MBH are always composed of the type of material opposite that of the parent proto-galaxy and because matter and antimatter gravitationally repel, therefore once formed MBH would not directly interact with material of the developing galaxy. Hence MBH would appear to be inert within the galaxy of creation yet their gravitational presence would still exist. Therefore, it is MBH that has been observed and recorded as "dark matter"—inert with regard to adjacent material yet emitting a strong gravitational presence that effects light passing nearby.

**Micelle** – A roughly spheroid structure produced as a result of self-assembly of a heterogeneous mixture of two opposing types of material. After self-assembly, the under-represented material would assume the micelle configuration while the predominant material would be the matrix surrounding the micelle.

**Micelle foam** – MBH that were carried to and deposited at the event horizon of supermassive central galactic black-holes (SCGBH) during the formation of these central structures. Being of opposite type material, micelle foam cannot interact with the SCGBH. They are held in static positions at the event horizon by equilibrium of opposing force exerted on them by both the SCGBH and the remainder of the galactic cluster. Micelle foam similarly exerts a repulsion force on all material within the remainder of the galactic cluster effectively barring the entry of new material into the SCGBH. In galaxies with a sufficient amount of micelle foam surrounding the SCGBH, no new material (or very little) would be expected to fall into the SCGBH. However, in galaxies with lower amounts of micelle foam, the SCGBH would remain active in feeding on adjacent material and growth. Thus accounting for why some observed SCGBH appear to be dormant while others continue feeding off their galactic cluster to this day. (See Conclusion 17)

**MPP (micellular positron packets)** – Segments within the solar wind matrix that because of primary gravitational sorting are composed solely of antimatter positrons. Because of gravitational repulsion with the star of creation, MPP travel away from the star's center and are the driver of the solar wind. Although MPP are immiscible with, and therefore do not interact directly with, solar material, they do produce a nudging on the diffuse material on the outer layers of the solar surface that ultimately leads to ejection of like-type material (in addition to MPP) from the solar surface.

In an antimatter-based galaxy, where there are antimatter stars being fueled by antihydrogen undergoing antifusion, and analog version of MPP would exist comprised of gravitationally sorted electrons (call them MEP.)

**Pair-production** – An event where energy is converted into mass as described by Einstein's equation, $E=mc^2$. In this event, two particles (one matter and one antimatter) are always produced and the net charge is always zero. If one particle is positively charged, the other will be negatively charged; or two neutrally charged particles could be produced.

**Positron** – The antimatter mirror equivalent of an electron. Essentially both possess the same mass and the same amount of electrical charge; however, an electron is negatively charged while a positron is positively charged. Positrons are "common" within the local Universe as they are a fundamental byproduct of fusion. Literally, tons of positrons are produced by our Sun every second.

**Self-assembly** – A naturally occurring phenomenon where an extremely heterogeneous system moves to a lower level of entropy in order to achieve greater stability. This process is the primary driver of immiscibility between matter and antimatter during the earliest moments of the Big Bang and is also observed within the heliospheric matrix during the production of MPP.

The supersymmetric analog of self-assemble within electrical systems has been documented repeatedly at the nano-level through observations of block copolymers and hydrophilic vs. hydrophobic interactions.

Organizational structures that are produced as a result of self-assembly include micelles, tubules, and layering depending on the types of materials used and their relative proportional representation. (See Conclusion 5)

**SCGBH (Supermassive central galactic black-hole)** – A structure formed during the Dark Event due to the strong attractive gravitationally like-like forces at the center of dominia. This imbalance caused the production/compacting into a black-hole. Formation of the SCGBH was stopped/slowed by the accumulation of micelle foam at its event horizon. Being incredibly stable, SCGBH persist at the centers of all galaxies through to the present day. (See Conclusions 10,17,18,19, and 20)

**Singularity** – This term is never used within the Dominium model. It is included within this glossary because it is a term that is commonly used semi synonymously with that of a black-hole and is derived from a conflicting model based on relativistic maths. To avoid unnecessary confusion and a desire not to refer to the relativistic methods of other constructs it is therefore not used.

**Solar wind** – The measured unidirectional flow of material away from the solar surface. Though this phenomenon has been verified for only a number of stars, it is theorized that all/most stars generate a solar wind.

Previous theories credit "the vacuum of space" for generating this observed phenomenon. However, no model has been put forward that can account for this or for the cyclical ebbs and flows that have been measured within this structure.

The Dominium model describes the solar wind as being the result of positrons (antimatter) being generated within the star as a result of fusion. Subsequently the matter of the star repulses them, and as these positrons attempt to leave the proximity of the star, they push stellar material outward as well. (See Part 3, which begins page 42)

**Supersymmetric equivalents** – Two different systems (one electrically based and one gravitationally based) that respond similarly to analogous types of conditions. For example, self-assembly of block copolymers is a supersymmetric equivalent to self-assembly of the original Universe matrix into immiscible dominia. Unlike previous considerations, the Dominium model considers the notion of supersymmetry at a systems level rather than at a formulaic particle level.

**Tunguska** – A region of Siberian forest that was flattened in 1906. At the time the area was uninhabited and there were no credible witnesses. Theories to account for the flattened forest range from aliens to freak weather. Almost half a century after the event, a research team went to the region to try to prove the hypothesis that a meteorite caused the destruction. They did "succeed" in finding fragments of extraterrestrial material, however, this finding is inconclusive because the Earth is constantly bombarded by extraterrestrial material and such fragments can be found everywhere across the surface of the globe. Also, the meteor-theory does not smoothly explain the pattern of fallen trees that appear to have been "pushed" down in the absence of an impact crater. Also, if a large meteor did enter the atmosphere, produce a forest-leveling shockwave, and then depart—such a meteor should still exist within the Solar System in an orbit that brings it close to Earth. However, none of the known near-Earth satellites possess trajectories that could be back calculated to have been passing close to Earth in 1906.

The Dominium explanation for this event is of a small antimatter MBH entering the Earth's atmosphere and subsequently being repulsed by gravitational interaction. (See Conclusion 40)

## Bibliography:


Abdullah, K; Haarsma, L; Gabrielse, G. Electronic accumulation of extremely cold positron in ultrahigh vacuum (1995) **Vol T59** pg 337-340

Anonymous. Antimatter's hidden shadow. (<2000) *Particle Physics and Astronomy Research Council*, Wiltshire UK

Anonymous. Big bang. (<2000) *Particle Physics and Astronomy Research Council*, Wiltshire UK

Anonymous. Edge of knowledge, the physics of the universe. (<2000) *Particle Physics and Astronomy Research Council*, Wiltshire UK

Anonymous. Mystery of matter and antimatter. (<2000) *Particle Physics and Astronomy Research Council*, Wiltshire UK

Aspinwall, P. Black-hole entropy, marginal stability and mirror symmetry. (2007) *Institute of Physics Publishing*

Bardeen, J.M; Steinhardt, P.J; and Turner, M.S. Spontaneous creation of almost scale-free density perturbations in an inflationary Universe (1983) *Phys Rev D* **Vol 28** pg 679-693

Bleicher, M. How to create black-hole on Earth (2007) *Eur. J. Phys* **Vol 28** pg 509-516

Bromley, M.W; Lima, M.A.P; Laricchia, G. On the XIIIth international workshop on low-energy $e^+$ and Ps physics (2006) *Phys Scr* **2006** C37-C45

Casado,R; Germani, C. Gravitational collapse and evolution of holographic black-holes (2006) *J.of Physics Conference Series* **Vol 33** pg 434-439

Cheng, J.Y; Ross, C.A; Smith, H.I; Thomas, E.L. Templated self-assembly of block copolymers: top-down helps bottom-up (2005) *Advanced Materials* **DOI:10.1002/adma.200502651** pg 2505-2521

Copi, I.M. Introduction to Logic 6th ed. (1982) MacMillan

De Bernardis; et. al A flat Universe from high-resolution maps of the cosmic microwaves background radiation. (2000) *Nature* **Vol 404** 27

Gao, S; Jiang, B.W; Zhao, Y.H. Possible streams of the globular clusters in the galaxy. (2007) *Chin. J. Astron Astrophys*. **Vol 7 No 1**, pg 111-119

Guth, A. Eternal inflation and its implications. (2007) *J.Phys.A.Math Theor* **Vol 40** pg 6811-6826

He, J.S; Tu, C.Y; Marsch, E. Can the solar wind originate from a quiet Sun region? *Astronomy & Astrophysics* **Vol 468, No 1**

Holzcheiter, M.H; Charlton, M. Ultra-low energy antihydrogen. (1999) *Rep Prog. Phsy* **Vol 62** pg 1-60

Karar, A; Musienko, Y; Vanel, J.C. Characterization of avalanche photodiodes for calorimetry applications *Nuclear Instruments and Methods in Physics Research: Elsevier Science* (1999) **A 429** 413-431

Kellerbauer, Alban; Walz, J. A novel cooling scheme for antiprotons. (2006) *New Journal of Physics*

Koch, B; Bleicher, M; Stocker, H. Black-holes at LHC? (2007) J. Phys G: Nucl. Part. Phys. **Vol 34** pg S535-S542

Kuhn, T. The Structure of Scientific Revolutions University of Chicago Press 1970 `

Lin, R.P, et.al. Three-dimensional plasma and energetic particle investigation for the WIND spacecraft (1995) *Space Science Reviews* **Vol 71** pg 125-154

MACHO Clues to Missing Mass (2000) *Frontiers: UK Particle Physics, Astronomy, and Space Science* **Issue 7** pg 8

McComas, D.J; Bame,S.J. Barker, P; Feldman, W.C; Phillips; Riley P. Solar wind electron proton alpha monitor (SWEPAM) for the advanced composition explorer (1998) *Space Science Reviews* **Vol 86** pg 563-612

Mewaldt, R.A, et.al. On the differences in composition between solar energetic particles and solar wind *Space Science Reviews* **DOI 10.1007/s11214-007-9187-1**

Miller, I.A, Instructor's Solutions Manual, Giancoli Physics *Prentice Hall* **5th ed.** Pg 231

MIT lecture – the words quoted were stated on June 26th, 2007 by an unnamed professor

Muller-Meskamp, L, et.al. Self assembly of mixed monolayer of mercaptoundecylferrocene and undecanethiol studied by STM (2007) *Journal of Physics: Conference Series* **Vol 61** pg 852-855

NASA-A public outreach module: solar wind, genesis, and the planets What is solar wind. http://genesismission.jpl.nasa.gov/science/module4_solarmax/SolarWind.html

NASA. When galaxies collide: chandra observes titanic merger (2002) **News Release #02-094**

NASA. Stars form surprisingly close to Milky Way's black-hole (2005) **News Release #05-166**

NASA ON the EXPANSION of the UNIVERSE from www.nasa.gov

*NASA ACE Program website[10]

NASA press release. http://science.nasa.gov/newhome/headlines/ast13dec99_1.htm

NASA public information website, www-spof.gsfc.nasa.gov/Education/wsolwind.htm

*NASA – Satelite specs – from nasa.gov

Ogilvie, K.W, et.al. SWE, a comprehensive plasma instrument for the WIND spacecraft (1995) *Space Science Reviews* **Vol 71** pg 55-77

Peroomian, V; El-Alaoui, M; Abdalla, M.A; Zelenyi. A comparison of solar wind and ionospheric plasma contributions to the september 24-25, 1998 magnetic storm (2007) *Journal of Atmospheric and Solar-Terrestrial Physics* **Vol 69** pg 212-222

Poth, H Atomic antimatter (1988) *Physica Scripta* **Vol 38** pg 806-815

[10] The asterisk indicates websites sourced in the research, yet where the http location not properly recorded. For the sake of time and the current stakes, publication of this work proceeds despite the missing data.

Riazantseva, M.O; Khabarova, O.V; Zastenker, G.N; Richardson, J.D. Sharp boundaries of solar wind plasma structures and an analysis of their pressure balance (2005) *Cosmic Research* **Vol 43, No 3** pp 157-164

Solar and Heliospheric Research Group http://solar-heliospheric.engin.umich.edu/science.htm

*Steele Hill/NASA Sun as Art not dated … downloaded from NASA.gov

Sueoka, Osamu Detection Efficiency of a Continuous Dynode Electron Multiplier (Ceratron) for Positrons (1982) *Japanese Journal of Applied Physics* **Vol 21 Issue 5** pg 702

Suess, Dr. Horton hears a Who (1954) *Random House*

*SWEPAM homepage

Temmer, M; Vrsnak, B; Veronig, A. Periodic appearance of coronal holes and the related variation of solar wind parameters (2007) *Solar Physics* **Vol 241** pg 371-383

Walz, J. Cold antihydrogen atoms. (2004) *Physica Scripta* **Vol 70** pg C30-34

Wignall, G. Neutron scattering studies of polymers in supercritical carbon dioxide (1999) *J Phys: Condens. Matter* **Vol 11** pg R157-R177

Printed in the United Kingdom
by Lightning Source UK Ltd.
136189UK00002B/92/P

9 780980 096323